THE LEARNING AND TEACHING OF MATHEMATICAL MODELLING

This book takes stock of the state of affairs of the teaching and learning of mathematical modelling with regard to research, development and practice. It provides a conceptual framework for mathematical modelling in mathematics education at all education levels, as well as the background and resources for teachers to acquire the knowledge and competencies that will allow them to successfully include modelling in their teaching, with an emphasis on the secondary school level. Mathematics teachers, mathematics education researchers and developers will benefit from this book.

Expertly written and researched, this book includes a comprehensive overview of research results in the field, an exposition of the educational goals associated with modelling, the essential components of modelling competency and an extensive discussion of didacticopedagogical challenges in modelling. Moreover, it offers a wide variety of illuminating cases and best-practice examples in addition to insights into the focal points for future research and practice.

The Learning and Teaching of Mathematical Modelling is an invaluable resource for teachers, researchers, textbook authors, secondary school mathematics teachers, undergraduate and graduate students of mathematics as well as student teachers.

Mogens Niss is Professor of Mathematics and Mathematics Education at Roskilde University (RUC), Denmark.

Werner Blum is Professor of Mathematics Education at the University of Kassel, Germany.

IMPACT: Interweaving Mathematics Pedagogy and Content for Teaching

The Learning and Teaching of Mathematical Modelling
Mogens Niss and Werner Blum

The Learning and Teaching of Geometry in Secondary Schools
A Modeling Perspective
Pat Herbst, Taro Fujita, Stefan Halverscheid and Michael Weiss

The Learning and Teaching of Algebra
Ideas, Insights and Activities
Abraham Arcavi, Paul Drijvers and Kaye Stacey

THE LEARNING AND TEACHING OF MATHEMATICAL MODELLING

Mogens Niss and Werner Blum

Routledge
Taylor & Francis Group
LONDON AND NEW YORK

First published 2020
by Routledge
2 Park Square, Milton Park, Abingdon, Oxon OX14 4RN

and by Routledge
52 Vanderbilt Avenue, New York, NY 10017

Routledge is an imprint of the Taylor & Francis Group, an informa business

© 2020 Mogens Niss and Werner Blum

The right of Mogens Niss and Werner Blum to be identified as authors of this work has been asserted by them in accordance with sections 77 and 78 of the Copyright, Designs and Patents Act 1988.

All rights reserved. No part of this book may be reprinted or reproduced or utilised in any form or by any electronic, mechanical, or other means, now known or hereafter invented, including photocopying and recording, or in any information storage or retrieval system, without permission in writing from the publishers.

Trademark notice: Product or corporate names may be trademarks or registered trademarks, and are used only for identification and explanation without intent to infringe.

British Library Cataloguing-in-Publication Data
A catalogue record for this book is available from the British Library

Library of Congress Cataloging-in-Publication Data
A catalog record for this book has been requested

ISBN: 978-1-138-73067-0 (hbk)
ISBN: 978-1-138-73070-0 (pbk)
ISBN: 978-1-315-18931-4 (ebk)

Typeset in Bembo and Stone Sans
by Apex CoVantage, LLC

CONTENTS

Series foreword vi
Acknowledgements viii

1 Introduction 1

2 Conceptual and theoretical framework – models and modelling: what and why? 6

3 Modelling examples 35

4 Modelling competency and modelling competencies 76

5 Challenges for the implementation of mathematical modelling 90

6 What we know from empirical research – selected findings on the teaching and learning of mathematical modelling 111

7 Cases of mathematical modelling from educational practices 145

8 Focal points for the future 189

Index *195*

SERIES FOREWORD

IMPACT, an acronym for *Interweaving Mathematics Pedagogy and Content for Teaching*, is a series of textbooks dedicated to mathematics education and suitable for teacher education. The leading principle of the series is the integration of mathematics content with topics from research on mathematics learning and teaching. Elements from the history and the philosophy of mathematics, as well as curricular issues, are integrated as appropriate.

In mathematics, there are many textbook series representing internationally accepted canonical curricula, but such a series has so far been lacking in mathematics education. It is the intention of IMPACT to fill this gap.

The books in the series will focus on fundamental conceptual understanding of the central ideas and relationships, while often compromising on the breadth of coverage. These central ideas and relationships will serve as organizers for the structure of each book. Beyond being an integrated presentation of the central ideas of mathematics and its learning and teaching, the volumes will serve as guides to further resources.

We are proud to present a book on the topic of *modelling* by two renowned authors. What prompted us to address this area? Only a few years ago, Felix Klein's third volume in the *Elementarmathematik series* – having been published in 1902 for the first time – appeared in English translation. This volume is – roughly speaking – devoted to "applied mathematics". Thus, applied mathematics should be an integral part of school mathematics, as Klein claimed one hundred years ago.

Working in mathematics is inextricably linked with translating from real-world contexts into mathematics and back. This is what "Mathematical Modelling" means, and thus modelling is significant for modern mathematics. Modelling enhances students' mathematical understanding of reality and supports their learning of mathematical process competencies.

Series editors
Tommy Dreyfus (Israel), Nathalie M. Sinclair (Canada) and Günter Törner (Germany)

Series Advisory Board
Abraham Arcavi (Israel), Michèle Artigue (France), Jo Boaler (USA), Hugh Burkhardt (Great Britain), Willi Dörfler (Austria), Koeno Gravemeijer (The Netherlands), Angel Gutiérrez (Spain), Gabriele Kaiser (Germany), Carolyn Kieran (Canada), Jeremy Kilpatrick (USA), Jürg Kramer (Germany), Frank K. Lester (USA), Fou-Lai Lin (Republic of China Taiwan), John Monaghan (Great Britain/ Norway), Mogens Niss (Denmark), Alan H. Schoenfeld (USA), Peter Sullivan (Australia), Michael 0. Thomas (New Zealand) and Patrick W. Thompson (USA)

ACKNOWLEDGEMENTS

The authors would like to thank the editors of the IMPACT series, Günter Törner, Tommy Dreyfus and Nathalie Sinclair, for their continuing support and their constructive patience as we worked on this lengthy project. We also want to thank Hugh Burkhardt and Caroline Yoon for their very helpful, encouraging and constructive comments and suggestions concerning the draft manuscript, and Martin Niss for his careful reading of the revised manuscript and his ensuing suggestions. Finally, we want to thank numerous colleagues who have provided significant advice and information to support our writing of the book; no one named, no one forgotten. Last, but certainly not least, we want to thank our wives, Elke Blum and Kirsten Niss, for their infinite patience with us during the extended and sometimes dense periods of hard work over the last several years.

1
INTRODUCTION

1.1 Prologue

Mathematics has been around for at least five thousand years. Throughout its existence, mathematics has been applied to deal with a host of issues, situations and phenomena outside of mathematics itself. This fact is reflected in the five-fold nature of mathematics (Niss, 1994): Mathematics is a *fundamental science* that deals with its own internally generated issues; it is an *applied science* that addresses problems and questions in scientific disciplines other than mathematics; it is a *system of instruments for practice* in culture and society; it is a *field of aesthetic expression and experience*; and it is an *educational subject* with a multitude of different manifestations that, in various ways, reflect the other four facets of mathematics.

This means that mathematics as a discipline never lived in "splendid isolation" from the surrounding world. On the contrary, there have always been intimate connections between mathematics and other disciplines and fields of practice – oftentimes collectively called "extra-mathematical domains". When mathematics and one or more of these domains meet, the encounter must involve both mathematics and the domain(s); neither side can be discarded. Sometimes the encounter is easy and straightforward, e.g., when only counting and elementary arithmetic are involved. Sometimes it is highly complex and difficult, as when a sophisticated mathematical theory is brought to bear on a new domain for the first time like for example in the theory of general relativity.

The purpose of involving mathematics in dealing with situations belonging to extra-mathematical domains is to help answer questions that arise in such situations. Perhaps the questions simply cannot be answered without the use of mathematics. Perhaps they can be answered in a better, faster or easier manner by way of mathematics. Invoking and activating mathematics to deal with a situation in an extra-mathematical domain (which for brevity will simply be called a "context" when appropriate) necessarily happens via an explicit or implicit construction of a mathematical model. Constructing such a model – or, differently put, undertaking *mathematical modelling* – consists of representing the main elements of a context with mathematical entities and the questions pertaining to the context with mathematical questions. The whole enterprise then consists of seeking answers to mathematical questions and interpreting these answers in terms of the context.

If the application of mathematics to other areas or disciplines is brought about by mathematical modelling, the people using mathematics in such contexts must be able to undertake mathematical modelling. Since an increasing proportion of the population will use mathematics in such contexts, the education system must equip learners with this ability; contributing to making this happen is the driving force behind this book.

1.2 What is this book about?

This book, which forms part of the IMPACT series, is about the learning and teaching of mathematical modelling. It has been written based on the following four observations, most of which will be further explained and elaborated on in various chapters of the book.

The first observation is that all students must learn to put mathematics to use in a wide variety of contexts. It is essential for societies and their citizens that these contexts include everyday practical and leisurely life with family and friends, occupational and professional life in the work force, the civic and societal life of citizens being concerned with culture and society, as well as life in specialised professional fields and academic disciplines that use mathematics to a significant extent. This is the main reason why mathematics is, by far, the world's most taught subject, measured by the number of students who study it in primary, secondary or tertiary education or by the number of lesson hours mathematics is taught. Since mathematics is used outside of the discipline itself by way of mathematical models and mathematical modelling, the education system must enable students to work with mathematical models and to undertake mathematical modelling.

The second observation is that mathematical modelling is, cognitively speaking, a difficult and demanding enterprise. Knowledge of and skills in pure mathematics, even if very solid and well founded, are not sufficient prerequisites for students to be(come) able to engage effectively and successfully in modelling activities. Much more is needed. In this book, this observation is further investigated and explained by a broad array of theoretical and empirical considerations.

The third observation is that mathematical modelling can, in fact, be successfully taught to and learnt by students. This requires teaching and learning to take place in environments that are diverse, multi-faceted and activity rich. For that to happen, we need teachers who are mathematically, didactically and pedagogically competent and committed. In particular, teachers need to be well versed in the teaching and learning of mathematical modelling themselves. This observation, too, will be further discussed in theoretical and empirical terms in this book, with particular regard to how the conditions for successful teaching and learning of mathematical modelling can be met.

The fourth and final observation is that, in spite of the previous observation, there is still a variety of strong barriers and challenges to be overcome if we want to ensure models and modelling an appropriate place and role in the teaching and learning of mathematics in ordinary classrooms throughout the world. We can observe, though, that more recent mathematics curricula and standards in several countries (e.g., the USA, Germany, Singapore, China, and Chile, to name only a few) include modelling as a compulsory component. Nevertheless, these barriers and challenges call for further investment of mental and material resources in research and development on teaching and learning mathematical modelling at all education levels.

Against this background, the primary focus of this book will be the teaching and learning of modelling, whereas the teaching and learning of given models and applications of mathematics will be of derived or secondary importance. Our overarching ambition with this book is to provide an up-to-date outline of the state of modelling in mathematics education and to indicate how modelling can be used in educational settings, with an emphasis on the secondary school level. We will take stock of the progress made in research and development, with a glance at selected educational practices in the field. One may describe this book as a research-based introduction to the didactics of mathematical modelling for interested parties who (are to) teach mathematical modelling at schools or universities, including teacher training institutions, or want to engage in professional development activities. It aims to provide important insights into the learning and teaching of mathematical modelling.

The structure of the book is the following. Chapter 2 is devoted to setting the stage of mathematical models and modelling by offering a detailed general conceptual and theoretical framework of the fundamental notions involved in this area of study. The chapter also deals with the cognitive aspects of mathematical modelling and with the role of models and modelling in the education system. In order to to provide flesh and blood to the largely theoretical exposition in Chapter 2, in Chapter 3, we present several modelling examples as a varied source of instantiations of our considerations and expositions in the chapters to come. Even though the examples are presented without specific regard to their actual or potential role in the teaching and learning of mathematics, each of them is accessible to lower or upper secondary students and, of course, to their teachers. Moreover, all of them have, in some form or another, been used in real classrooms. The concepts of modelling competency and (sub-)competencies are introduced and discussed in Chapter 4. Chapter 5 focuses on the challenges and barriers to the inclusion of mathematical modelling in mathematics education that have been encountered in different places. These must be overcome if mathematical modelling is to become a substantive component of the teaching and learning of mathematics. While multiple references to theoretical and empirical research are found in all chapters of the book, Chapter 6 presents a comprehensive survey of empirical research on a variety of key aspects of mathematical modelling in mathematics education. Research and development in the didactics of mathematical modelling are the main focus of this book. However, educational practices of mathematical modelling at different levels must not be excluded from this book, especially since it is still an unusual – if not outright esoteric – topic in many mathematics curricula around the world. Chapter 7 is therefore devoted to presenting a few selected cases of different states of implementation of mathematical modelling in the teaching and learning of mathematics. Finally, Chapter 8 attempts, in a more global way, to take stock of what we know and have accomplished in the field of didactics of mathematical modelling, providing a point of departure for looking into future needs and challenges.

By its very nature, this book is, in large part, an exposition of what already exists and what has been done in the field of mathematical modelling, by others as well as by ourselves. However, in several places, we offer conceptualisations, approaches and perspectives that are new to the field. In addition, we include some novel views on the field. We realise that these are *our* views and that others working in this field may prefer other takes.

A final remark on the nature of this book in the fauna of other books about mathematics education is warranted. The didactics of mathematical modelling differ from many other sub-fields of mathematics education in that it is not entirely subsumed under the umbrella of mathematics. Extra-mathematical needs, demands, facts, aspects and elements are necessarily present in crucial ways in any kind of mathematical modelling, even in its highly idealised and stylised manifestations. One consequence of this is that the teaching and learning of mathematical modelling unavoidably must transgress the borders of mathematics to move into domains ruled by other sorts of preoccupations, forms of knowledge and methodologies than the ones characteristic of mathematics. Therefore, mathematical modelling cannot be addressed solely by mathematical means, so students and teachers engaging in mathematical modelling must locate, adopt and activate knowledge outside of mathematics as a prerequisite to or a part of their modelling work. For many a student or teacher, this presents major challenges.

1.3 What this book is not about

In mathematics education research and development, it is not unusual to invoke the term "modelling" in several ways, many of which have nothing to do with mathematical modelling as the term is understood and used in this book.

The most widespread alternative use of the term "modelling" is encountered in attempts to model students' mathematical thinking, mathematical problem solving, mathematical mistakes, mathematical behaviour, etc. In such attempts, "modelling" means establishing a conceptual and theoretical framework designed to make sense of students' engagement with mathematics by interpreting their actions, behaviour and statements. Only in the special case of "modelling students' mathematical modelling" (see Lesh et al., 2010) is this notion of modelling relevant to this book. In the theory of cognitive apprenticeship (see, e.g., Brown et al., 1989), the first of several phases is also called "modelling". In this phase, the teacher demonstrates, as an expert (a role model), how to tackle a typical task in the topic area that the students are to address. If this task is a modelling task, this means the teacher serves as a model of how to model a situation. Again, this use of "model" and "modelling" according to cognitive apprenticeship is not what we are dealing with in this book.

Quite a different notion of "modelling" is found outside of mathematics education – in the logical foundations of mathematics and in mathematical logic as a separate topic – in which a concrete mathematical theory is perceived as a realisation – a model – of some abstract axiom system. From this perspective, mathematical modelling is the activity of identifying mathematical entities which display the properties of a given axiom system. The present book does not adopt this perspective.

Similarly, it often happens in work within the discipline of mathematics that one mathematical theory, say linear algebra (or general topology), is introduced to model another mathematical theory, say Euclidean geometry (or real analysis). While such (intra-mathematical) modelling is indeed both very important and highly illuminating, this notion of modelling does not form part of this book. A somewhat analogous situation is found with regard to the notions of horizontal and vertical mathematisation, initially proposed by Treffers in 1978 (Treffers, 1993) and later supported by Freudenthal (1991). Horizontal mathematisation is the process of building a mathematical model of some situation outside of mathematics, which is what we call mathematical modelling in this book. Vertical mathematisation, in contrast, is the process of subjecting a problem formulated within mathematics to internal mathematical treatment in order to solve the problem. In the context of mathematics education, vertical mathematisation is typically introduced once a horizontal mathematisation has been undertaken. In this book, we have not adopted the notion of vertical mathematisation as this term is at odds with key notions in our exposition of mathematical modelling.

References

Brown, J.S., Collins, A. & Duguid, P. (1989). Situated cognition and the culture of learning. In: *Educational Researcher* **18**, 32–42.

Freudenthal, H. (1991). *Revisiting Mathematics Education: China Lectures.* Dordrecht: Kluwer Academic Publishers.

Lesh, R., Galbraith, P.L., Haines, C.R. & Hurford, A. (Eds.) (2010). *Modeling Students' Mathematical Modeling Competencies: ICTMA 13.* New York, NY: Springer.

Niss, M. (1994). Mathematics in society. In: R. Biehler, R.W. Scholz, R. Strässer & B. Winkelmann (Eds.), *Didactics of Mathematics as a Scientific Discipline* (pp. 367–378). Dordrecht: Kluwer Academic Publishers.

Treffers, A. (1993). Wiskobas and Freudenthal: Realistic mathematics education. In: *Educational Studies in Mathematics* **25**, 85–108.

2

CONCEPTUAL AND THEORETICAL FRAMEWORK – MODELS AND MODELLING

What and why?

2.1 Basic mathematical models – models as sheer representation

For centuries, mathematics has been used for multiple purposes, and in lots of different ways, in a wide variety of extra-mathematical domains, i.e., areas outside of mathematics itself. Extra-mathematical domains can be other academic disciplines or professional fields; they can be vocations, professions or other areas of practice; they can belong to societal and social spheres; or they can be part of everyday life with families and friends. The very point of involving mathematics in such contexts is that mathematics is expected to be useful for dealing with situations arising in these contexts. The purpose of this involvement may either be to come to grips with certain already existing aspects of the context and domain at issue, or it may be to design new elements, systems or features for implementation within the domain. We shall say much more about this in section 2.5.

What do we mean by a mathematical model and by mathematical modelling?

Every time mathematics is used outside of mathematics itself, a so-called *mathematical model* is necessarily involved, either explicitly or – very often - implicitly. But what is a mathematical model? Let us answer first a slightly more general question: What is a model? A model is an object (which is oftentimes in itself an aggregation of objects), which is meant to stand for – to represent - something else. The model is meant to capture only certain features of the entity it stands for and is thus a simplified representation of this entity. This simplified representation necessarily – and intentionally – involves some loss of information, hopefully information of less significance in the context at issue.

Simply put, a mathematical model is a special kind of model, namely a representation of aspects of an extra-mathematical domain by means of some mathematical entities and relations between them. In its simplest possible form, the situation can be depicted by the following diagram (Figure 2.1):

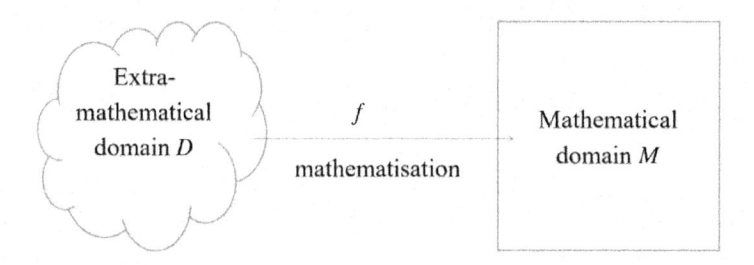

FIGURE 2.1 The minimal modelling diagram

Here, the "amoebic blob" on the left stands for the extra-mathematical domain, of which some selected aspects are to be represented by mathematical entities belonging to some chosen mathematical domain, M, depicted by the box on the right-hand side. The mathematical representation takes place by way of some kind of "mapping",[1] f, represented by the arrow between the boxes, which translates selected objects from D into selected objects from M. It is important to be aware that the selection of those objects in D that are to be given a mathematical representation is usually not at all an automatic process; it is a conscious act of will on the part of the modeller. The same is true of the choice of the domain M; the selection of the objects in M chosen to represent the objects in D; and the specific way in which the objects in D are linked to objects in M, as captured by the "mapping" f. The outcomes of all these selection and choice processes are designated in condensed form by the triple (D, f, M) which then may serve as a formal definition of a mathematical model (Niss, 1989). The fact that there are many choices and decisions involved in establishing the model makes it clear that establishing a model is indeed a process. This process is termed *mathematical modelling*. Often, several different combinations of choices and decisions might be taken into consideration to model a given situation. This means that a mathematical model only in trivial cases is uniquely determined by the situation. It often happens that people, when speaking about a mathematical model, refer only to M, or a subset thereof, as the mathematical model, without involving D or f. Defining a mathematical model as the triple (D, f, M) makes it clear that any mathematical model is a model of *something* in an extra-mathematical domain, not just a collection of mathematical entities, such as a certain function; a set of algebraic, functional or differential equations; or a planar or spatial geometric object, to mention just a few.

Even this simplest possible sort of model, which only consists of the sheer representation of extra-mathematical entities by mathematical ones, has a widespread use, even if the terms "model" and "modelling" are seldom used in such simple cases. One will sometimes speak about "coding" the set of objects instead of about modelling it. For example, in most cities in the world, buildings or main doors in buildings are equipped with street numbers that allow one to find the place where somebody lives or works without having to consult a lengthy and wordy description of the appearance of the building or door. As several people may have the same street number, you will usually not be able to identify a particular individual solely on the basis of his or her address. Therefore, a code need not be invertible. In this case, D consists of the collection of inhabitants on, say, a certain street, whereas M may be chosen to consist of the set N of natural numbers or some finite subset thereof.

Similarly, bar codes of, say, supermarket articles are commonplace all over the world. They are designed to identify the specific category to which a given article belongs and to derive information from this category, e.g., the current price of the article. Given the bar code, one can identify the category of the article but usually not the individual specimen within the category because this is not significant to the intended use of the bar code in supermarkets. So, this coding, too, is not invertible to the individual level. Here, D is chosen to consist of all the articles for sale in a supermarket. First, each article is coded by a finite string of base-10 digits which is specific to the category of the article. Let us denote the set of feasible finite strings of base-10 digits by M' and the mapping from D to M' by f'. This gives rise to the model (D, f', M'). Next, each digit is then coded by a short sequence of black and white line segments of varying thickness (i.e., very slim rectangles) such that the number code of the article is represented by a finite string of such line segments, the collection of which is denoted by M. There is a one-to-one correspondence – let's denote it by $g: M' \to M$ – between the finite strings of base-10 digits in M' and finite strings of black and white line segments in M. This means that a given bar code provides a geometric representation in M of any article in the supermarket and hence – because of the one-to-one correspondence between M and M' – a multi-digit number representation in base 10 of the article. The mathematical properties of the numbers are only used to construct a final control digit to check whether a given article code is a legal one. So, the bar code model is (D, f, M) if we set $f = g \circ f'$, i.e., define f as the composite of f' and g. The very point of using bar codes on articles instead of numbers is that the bar codes can be easily read with almost no errors by optical code scanners, whereas reading a string of digits, which may be written in a multitude of different ways, would be either much slower or more error prone.

Sometimes, sheer representation models are in fact invertible. For example, this is the case with car number plates in most countries (or states), where number plates contain alphanumeric codes establishing a one-to-one representation of all registered cars in the country (or state). As another example, many countries (e.g., Denmark and Sweden) have each of their citizens represented by a uniquely determined person identification number (a string of natural numbers in base 10), social security number or whatever the number is called in a particular country. The point of using such numbers to represent individuals is to obtain a compact, easily recordable one-to-one correspondence between individuals and numbers. The only "number feature" used in such a representation is that there is usually some arithmetic operation that has to be performed on the digits (e.g., multiplying each digit with a certain natural number and then adding the resulting numbers) to ensure that the number is legal according to the rules of the country at issue. No other mathematical operations are used on the identification numbers. Thus, adding or multiplying – say - identity numbers are meaningless operations even though they are, of course, technically possible. In principle, any system of establishing a one-to-one correspondence between the domain of interest and a domain of representation might be considered.

Invertible representation models can be captured by the following diagram (Figure 2.2):

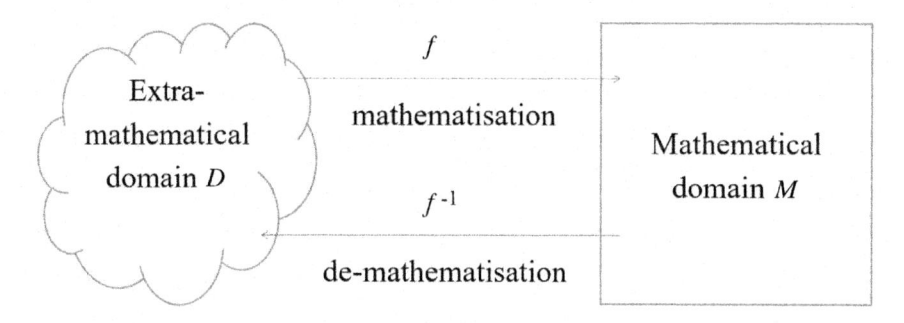

FIGURE 2.2 Minimal modelling with inversion

Here, it is possible to go from a particular code object in the relevant part of *M* back to a uniquely determined object in *D* that is coded by this code object. In general, going back from the mathematical domain *M* to the extra-mathematical domain *D* is called "de-mathematisation" (or interpretation and, in special cases, such as in the examples above, de-coding). Here, de-mathematisation takes place by simply inverting the mapping *f*.

2.2 Mathematical models with a structure

Normally, however, the terms "mathematical models" and "modelling" are primarily used in situations in which more structure than offered by sheer coding is either already present or is being requested. Despite this, it is analytically significant for the sake of the conceptual scope of our definitions to speak of mathematical models and modelling also in the minimal cases discussed in section 2.1. However, if we move into more complex contexts and situations that call for really utilising mathematical properties of the representing entities, more structure typically must be invoked or introduced. The call for more structure comes from a wish to represent certain properties of the relevant objects in *D*, as well as relationships between such objects by mathematical means. This, then, must be reflected in mathematical properties of the mathematical domain *M*, of the objects in *M* chosen to represent the objects in *D*, and of the mathematical relationships between those objects. For illustration, let us think of someone who wants to make a savings arrangement in a bank by making a fixed instalment every month for a set period and then wants a prediction of the accumulated fortune after a certain number of years. Answering this question by mathematical modelling requires, as a minimum, not just representation of instalments, terms, fixed or varying interest rates, sums, etc. It also requires the utilisation of mathematical properties of sums of number series, e.g., geometric series, relevant formulae, etc. (also see example 5 in Chapter 3).

The very reason we are interested in models with some mathematical properties is that such models often allow us to pose and *answer questions* concerning the extra-mathematical domain under consideration. Before moving on, let us illustrate the situation by means of a familiar and rather simple example. We consider modelling the cost of a taxi ride from S (starting point) to A (arrival point) in a certain city C. It is well known that taxi tariff systems vary from place to place and from one taxi company to another. All sorts of different conditions may apply concerning the size of the taxi, the time of the day chosen for the ride, the number of passengers, the kinds and amounts of luggage, whether the taxi is ordered by radio or hailed on the street, and whether it drives through pay roads or serves an airport. Also, the presence and size of a basic rate, the length of the ride or the waiting time at streetlights or in traffic jams are key factors in determining the cost of a ride. Oftentimes, much of this information is available from companies' home pages. For someone who wants to model the price of a taxi ride in C under various circumstances, the extra-mathematical domain, D, consists of the "universe" of taxi rides in C and their costs, and the modelling situation embedded in this domain is taxi rides from S to A. In principle, modelling such a ride would involve taking all sorts of factors and conditions, like the ones mentioned above, into consideration and would be a substantial and time-consuming task. The modeller, therefore, wants to first model a much-simplified situation (which might later be expanded or generalised) corresponding to the typical needs of a taxi passenger. Imagine that the modeller decides to focus on a taxi ride for one person without luggage in a car from one particular company during working hours (9am-5pm) and thus wants to answer the question: What does such a ride cost? To answer this question, the modeller makes some simplifying assumptions pertaining to the situation. For example, the modeller may assume that the taxi is called by radio; that the ride takes place along a particular route; and that the cost depends only on an initial fixed basic cost b and the variable distance travelled, x, and the rate d (the cost per unit distance), whereas waiting time, pay-zone fees, and other factors are left out of consideration at this point. More specifically, the modeller chooses to model the situation by the following linear[2] real function, c, of one variable:

$$c(x) = dx + b.$$

Here d, x and b are non-negative real numbers, and it is assumed that the distance component of the rate is an additive term proportional to the distance travelled. If the currency is DKK and the distance, x, travelled is measured in km, $c(x)$ and b are measured in DKK, d in DKK/km.

The mathematical domain, M, may be chosen to consist of the world of real linear functions of one variable. We may also choose a larger domain, for instance, the world of all real functions of one variable, the world of linear functions of several variables, or the world of all real functions of any number of variables.

In order to answer the question "What does a taxi ride cost?", we need to know the basic rate, the rate for 1 km, and the distance travelled. Consulting the website of one of the taxi companies in C, we learn that the basic rate is 38 DKK and the rate per km is 15.55 DKK, so the function is specified to be:

$c(x) = 15.55x + 38.$

According to available maps, the distance between S and A is 12.9 km along the stipulated route. So, we can now translate our real-world question into a corresponding mathematical question: What is the value $c(12.9)$?

Answering extra-mathematical questions by way of mathematical modelling is brought about by translating these questions into mathematical questions concerning the mathematical entities selected to represent the extra-mathematical entities in focus of our attention and then to seek *mathematical answers* to the translated questions. The answering of the mathematised questions takes place by performing mathematical processes, including calculations and computations, and by making mathematical inferences to obtain mathematical conclusions and results within the mathematical domain of the model. In the example above, the mathematical question is answered by simply performing the calculation 15.55 · 12.9 + 38, yielding 238.60. It is then expected – or at least hoped - that the mathematical answers obtained can be translated back – de-mathematised - into answers pertaining to the extra-mathematical domain being modelled. In our example, this amounts to attaching a unit (DKK) to the number obtained to formulate an answer: The taxi ride costs DKK 238.60.

We have just seen an example of what we may call *the fundamental mathematical modelling process involving structure*, which can be represented by the following diagram (Figure 2.3):

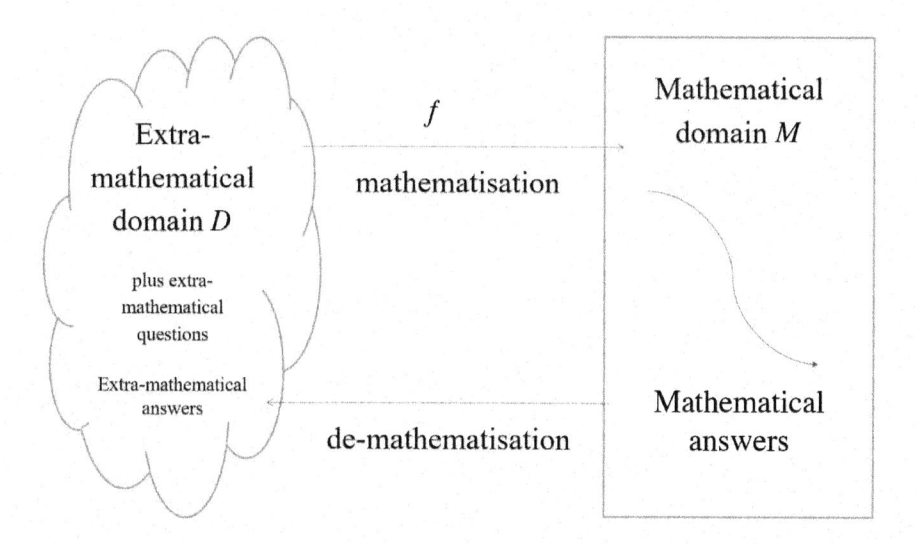

FIGURE 2.3 Diagram of fundamental mathematical modelling involving structure

In a given modelling situation, the fact that we might have been able to obtain mathematical answers to our mathematised questions and to de-mathematise these answers into extra-mathematical answers does not necessarily imply that the answers obtained meet the needs and demands that drove our modelling enterprise in the first place. To check whether this is the case, we must *validate the answers* of the model, i.e., examine the extent to which the answers obtained are relevant and useful for the purpose for which the model was constructed, in addition to being solid and well-founded. Do the answers correspond to and cover the needs associated with the initiating questions? Are the answers complete or only partial? Do they cover all relevant instances of the situation being modelled, or do they depend on special circumstances and conditions? How sensitive are the outcomes to the assumptions made, to the accuracy of available data, input variables, parameter values, etc.? In our taxi example, the answer obtained is not unreasonable. It gives an indication of the cost of a ride under the constraints involved. However, the answer does depend on special circumstances and conditions that easily might have been different. Other hours for the ride, other routes, or other taxi companies might have been considered. Moreover, waiting time at traffic lights and in queues, which is known to be significant, has not been considered. Therefore, the cost arrived at will be the minimum cost for such a ride.

Validating the answers produced by a model is but one instance of a wider issue: How good is the model? More broadly, we would like to *evaluate the model* (also see Czocher, 2018). Although the ultimate purpose of evaluating a model is to decide whether to accept or reject it, it is not a matter of deciding whether or not it is correct or incorrect but rather a matter of finding out how well it suits its purposes as we have defined them. In addition to validating the model answers, there are basically three additional ways to undertake an evaluation of the model. One is a qualitative assessment of the *structural properties* of the model – are the phenomena and features displayed by the model compatible with what is already known about the extra-mathematical reality? How robust is the structural behaviour of the model beyond the specific setting within which it was built? How dependent is it on the assumptions made? Can it be generalised or modified to cover a broader class of situations?

The second way is to assess the *quantitative features* of the model. This can consist in, for instance, confronting quantitative model outputs with known data – e.g., to see whether output values lie within reasonable or useful ranges – or display a satisfactory degree of accuracy, or whether the model is capable of capturing quantitative aspects of the extra-mathematical domain not directly represented in the model. Also, do statistical analyses, if carried out, lead to a fair degree of confirmation, e.g., in terms of statistical significance, of the conclusions? Moreover, it will be important to know whether the model is under-parametrised, which means that there are not enough data to allow for a complete specification of the model or for a satisfactory estimation of non-measurable parameters. Or whether – on the contrary – the model is over-parametrised, which means that a multitude of different parameter values can be generated based on the data available, thus giving rise to the model being indeterminate.

The third way to evaluate a model is to compare and contrast it with *possible alternative models* meant or supposed to cover the same extra-mathematical domains for similar or related purposes, especially models based on assumptions that differ from the ones made for the model under consideration. In the case of the taxi example, an obvious alternative – or perhaps extended – model to consider would be to include waiting time. The website of the company mentioned above tells us that the additional cost of 1 minute of waiting time is DKK 7.14. If we want to take this into account, we must introduce a new mathematisation, for instance:

$c'(x, t) = 15.55x + 7.14t + 38,$

with non-negative real numbers x and t, where the unit of the coefficient in front of t, which is measured in minutes, is DKK/minute. This mathematisation requires us to invoke a mathematical domain that is rich enough to include real linear functions of two variables. This change of model also leads to a change of the extra-mathematical questions we want to ask and to the mathematised versions of these questions.

2.3 The modelling cycle

By adding the processes of validating the model outputs and evaluating the model, we have dealt with the most essential components of mathematical modelling. Taken together, these components constitute the basic version of what is usually called the *modelling cycle*: mathematisation; mathematical treatment; de-mathematisation; and answer validation and model evaluation – depicted in Figure 2.4. It is essential

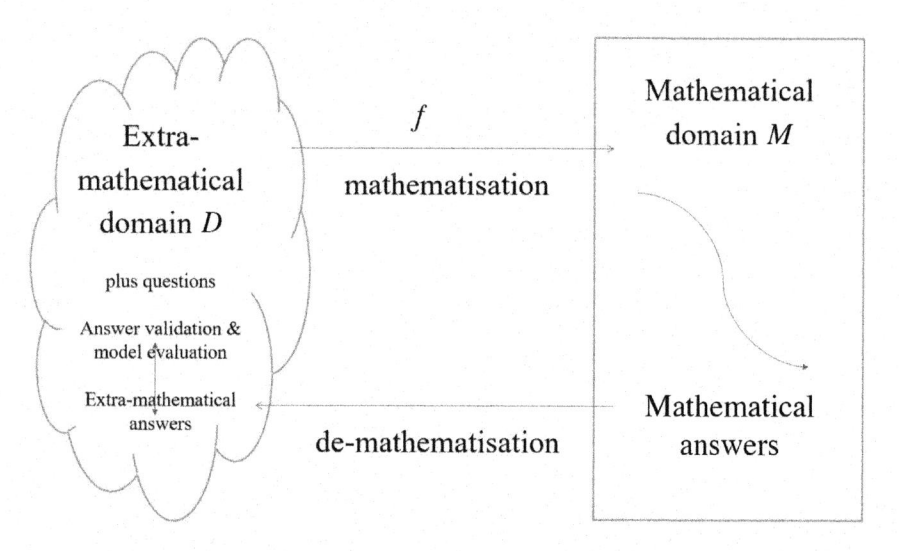

FIGURE 2.4 The basic modelling cycle

to understand that this modelling cycle is not meant to be a description of the sequence of actions that people must take, or do actually take, in the order listed, when building a model. People's actual modelling itineraries may take all sorts of forms, starting anywhere within or outside the cycle, repeating some sub-processes several times, skipping others, etc. (see section 6.3 for further details). The modelling cycle therefore should be understood as an *analytic* (ideal-typical) *reconstruction* of the steps of modelling necessarily present, explicitly or implicitly, as an instrument for capturing and understanding the principal processes of mathematical modelling.

Depending on which aspects of the key steps of the basic modelling cycle are of special interest in each context, the cycle can be expanded by zooming in on the details of these aspects. If, for instance, the focus is finding or choosing a situation worth further consideration within a large, not clearly delineated extra-mathematical domain, it is significant to look into ways to delimit that domain. Such delimitations must be compatible with the very purpose of dealing with the extra-mathematical domain. If, on the other hand, the focus is on the posing and specification of the extra-mathematical questions to be considered, it is essential to select the elements and features that are deemed important for the situation at issue and discard those that aren't. It is also essential to make extra-mathematical assumptions concerning the properties of the elements selected and the relationships between them. The role of all these processes, which take place within D, is to tailor the extra-mathematical domain in order to produce a reduced extra-mathematical situation with accompanying questions from it and to prepare that situation for translation into some mathematical domain, i.e., for *mathematisation*. We therefore call the collection of these processes *pre-mathematisation*. Essential components of pre-mathematisation include specifying the elements of the situation that should be considered, the features and properties to be considered, as well as known or assumed relationships between the elements, including the mechanisms that govern those relationships. As an example of pre-mathematisation, we may consider our taxi case presented in the previous section. Here, all the considerations involved in deciding which aspects and conditions to pay attention to and which to discard that led to a (much) reduced modelling situation and corresponding questions constitute the first part of pre-mathematisation. The remainder of the pre-mathematisation consists of all the assumptions made to specify the situation further.

The result of pre-mathematisation is a tailored, condensed, structured and possibly idealised extra-mathematical situation – *a reduced extra-mathematical situation cum questions* – containing exactly those demands, elements, factors, components, assumptions and relationships that should be subjected to mathematisation. Depending on the situation and the questions at issue, it is sometimes even possible to organise pre-mathematisation into an extra-mathematical model – sometimes called a *real model* (Blum, 1985) or a *real-world model* (Kaiser & Stender, 2013) – of the situation. It is important to keep in mind that even though pre-mathematisation is performed with a view to the subsequent process of mathematisation, it still belongs to the extra-mathematical world in which the context is embedded. The same is true of a possible extra-mathematical model, in case one is established.

Conceptual and theoretical framework **15**

One such expansion of the modelling cycle, zooming in further on pre-mathematisation, can be found in Niss (2010). A slightly edited version is given in Figure 2.5.

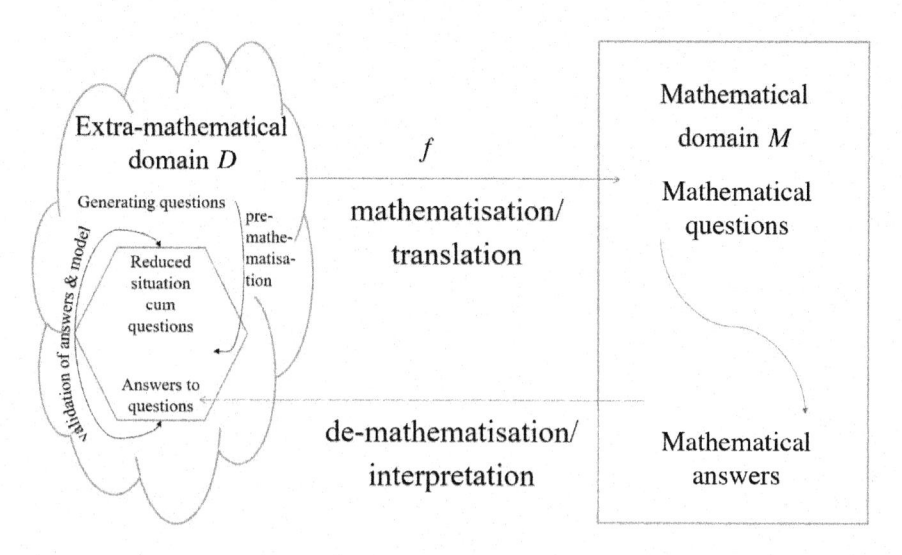

FIGURE 2.5 The modelling cycle

Another version of the modelling cycle can be found in Blomhøj & Jensen (2007) (see Figure 2.6.)

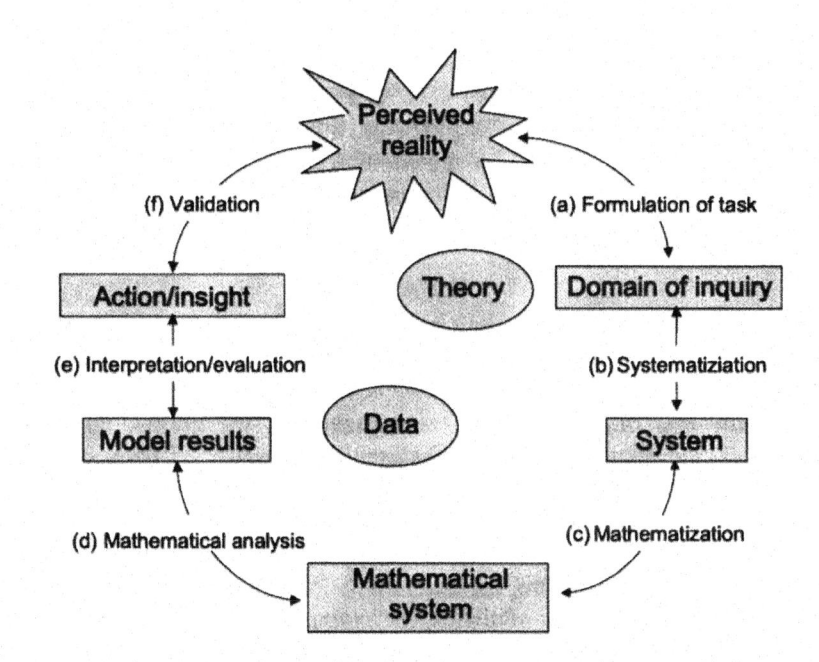

FIGURE 2.6 The Blomhøj-Jensen modelling cycle

16 Conceptual and theoretical framework

In this diagram, the steps preceding that of mathematisation focus on formulating a task inspired by the perceived reality, giving rise to a more specific domain of enquiry and further on (as a result of systematisation) to a more or less well-delineated system, which is then subjected to mathematisation. At the other end of the cycle, the model results, after interpretation and evaluation, generate actions or insights pertaining to the perceived reality. In Galbraith and Stillman (2006), we find the following version of the modelling cycle (Figure 2.7):

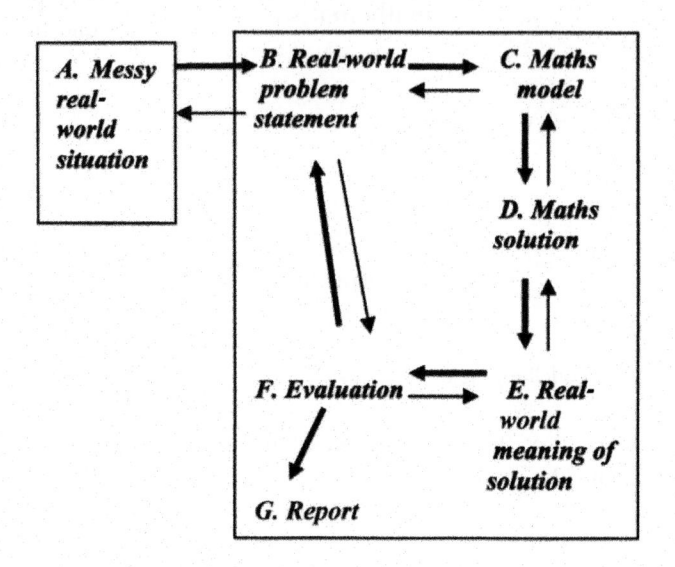

FIGURE 2.7 The Galbraith-Stillman modelling cycle

in which the "messy real-world situation" is only part of the modelling process insofar as it gives rise to a real-world problem statement. The other components of the modelling cycle closely resemble those in the modelling cycle diagrams presented above, except that here particular attention is being paid to an additional component, namely a report of the modelling work undertaken, which implies the existence of some kind of audience or recipient of this work.

The modelling cycle found in Blum and Leiss (2007) (see Figure 2.8) tacitly operates with the presence of a modelling task given to a modeller – suggesting a school context in which a teacher assigns tasks to students. The diagram pays particular attention to the modeller's need to understand the task before simplifying and structuring it into, first, a *situation model* – i.e., a mental image of the fundamental characteristics of the situation and its essential elements – and, second, into a real model, all of which are cognitive processes referring to the domain of reality. Finally, the flavour of a task set from outside is amplified by the presence of the last step in the cycle, "presenting" the work to an audience or a recipient, e.g., a class or a teacher.

Conceptual and theoretical framework 17

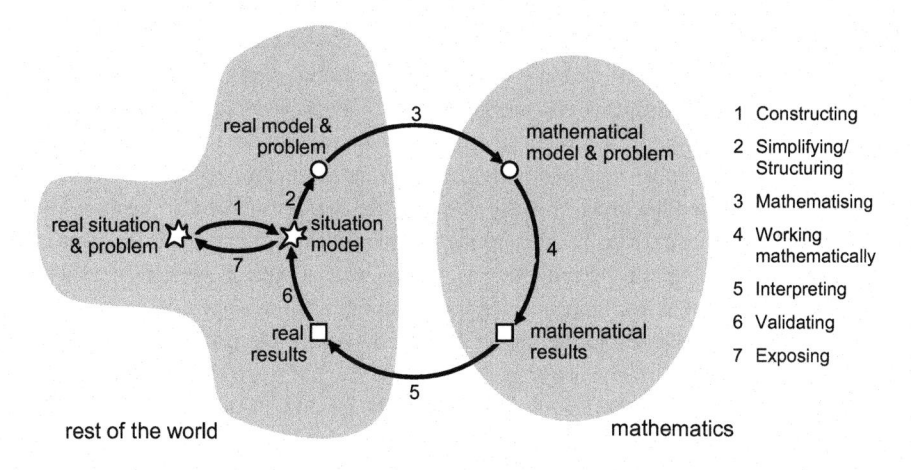

FIGURE 2.8 The Blum-Leiss modelling cycle

Many more modelling cycles can be found in the literature (for an overview, see Borromeo Ferri, 2006; Perrenet & Zwanefeld, 2012; or Blum, 2015). All the different "models of the modelling process" describe and visualise connections between an extra-mathematical world and mathematics, and they identify certain steps on the pathway from the extra-mathematical world into mathematics and back again.

As is clear from the above examples, some of the cycles have other purposes than describing the generic steps of modelling. For instance, they may focus on the cognitive processes of the modeller or on making a report or a presentation of the work done to some audience. If, in a given context, the focus of interest is the cognitive aspects of the process of mathematising, or obtaining, within the mathematical domain M, answers to the mathematical questions resulting from the mathematisation by way of mathematical problem solving, the modelling cycle can be expanded accordingly. The same is true if attention is being paid to the sub-processes involved in de-mathematisation or validation. We come back to the cognitive aspects of modelling in section 2.6.

2.4 Some special kinds of modelling

A special kind of modelling deserves particular attention when we discuss the modelling process. It is customary in a variety of extra-mathematical fields to engage in so-called *curve-fitting*, or more broadly *regression analysis*. Oftentimes a modeller is faced with a set of pairs of linked real-valued data, (x_1, y_1), (x_2, y_2), ..., (x_n, y_n), where it is assumed that there is some sort of a functional relationship between the x's and the y's, which can be used to make interpolations or extrapolations in between or beyond the data set, but the specific functional relationship is not known. The modeller then wants to come up with a functional relationship to mathematise the situation.

In case the modeller, based on knowledge about the extra-mathematical context and situation, has reasons to believe that the functional relationship is of a particular type – for example a trigonometric, linear or polynomial function, power, exponential or logarithmic function – the mathematisation step falls under what has already been said above. Specifying the exact specimen among the functions of the given type then becomes an issue of using the data set at hand to procure the parameter values needed to identify the specific function. This is what happens in the phase of mathematical treatment, which also may comprise statistical methods and techniques, including statistical regression analysis. In other words, in this case, nothing new is happening in the modelling process compared to what has been said in section 2.3.

However, in case the modeller has no clue of the function type of the data set, the mathematisation step takes a very different form. The modeller will have to impose some type of function onto the data set (typically after having entered the pairs as points in a coordinate system) without having chosen its type based on structural or substantive considerations pertaining to the extra-mathematical domain. On what grounds, then, can the modeller impose a function type on the data set? Either by being visually inspired by the location of the data points in the coordinate system – perhaps the points seem to approximately lie on a straight line; on an S-shaped curve; on a curve displaying regular, damped or increasing oscillations; on a steeply increasing or decreasing curve with or without apparent asymptotes, etc. – which suggests to consider a certain function type, or by making use of the fact that through n ($n > 1$) points in a coordinate system, no two of which have the same x-value, there is a uniquely determined polynomial of degree at most $n - 1$ whose graph contains the n points. Determining that polynomial and assuming that the polynomial also represents all potential data points in the relevant domain then provides a mathematisation of the situation and a model of it (see also Burkhardt, 2018, p. 62).

From a formal point of view, these two sorts of mathematisation are completely acceptable on a par with any other sort of mathematisation. But this modelling process is markedly truncated since no pre-mathematisation was involved in generating the mathematisation. In fact, one might perceive the mathematisation as resulting from a particular kind of mathematical treatment of the data set. This means that the first part of the modelling cycle is not relevant in this context. However, mathematical treatment, de-mathematisation, answer validation and model evaluation, the last phases of the modelling process, can be dealt with in the very same way as with any model obtained after first undertaking pre-mathematisation. Similarly, the corresponding parts of the modelling cycle also remain relevant in this situation.

Another notion to consider is that of *applied mathematical problem solving*. In the past, this notion was often used as a synonym for mathematical modelling, especially in the USA, or to designate – primarily in the UK – mathematical problem solving in the special context of theoretical mechanics or fluid dynamics, both branches of physics in which physical theories provide the framework under which modelling has to be subsumed and performed. Today, the notion of applied mathematical problem solving seems to have taken on a new meaning, namely to designate mathematical problem solving arising from problems pertaining to an extra-mathematical domain for which a mathematical model has already been established prior to the problem-solving activity. For example, consider a situation in which taxi company A charges $p_A(x,t) = 15.55x + 7.14t + 38$ for an x km long ride with waiting time t (see section 2.2), whereas company B charges $p_B(x,t) = 16x + 6t + 36$ for the same ride, and the question is: "Under what circumstances should company A, company B, be chosen?" Like in the case of curve-fitting, we are still dealing with a mathematical model but not with a full-fledged modelling process because the model has been given and both the modelling process and the modelling cycle are truncated. So unless the applied mathematical problem solver was also actually him- or herself responsible for constructing the model, we do not, in this book, consider applied mathematical problem solving to be part of mathematical modelling.

While curve-fitting and applied mathematical problem solving involve little or no pre-mathematisation and mathematisation, the so-called *Fermi problems* lie at the other end of the scale. Typical examples of Fermi problems are: "How many piano tuners are there in the city of Chicago?"; "What is the area covered by a litre of water spilled on the floor?"; "How many leaves of grass are there in a 1m^2 grass lawn?"; "How many people can be seated for dinner in a gym hall?"; "How many hairs does a girl have?", to mention just a few. While they are, in principle, closed when it comes to the kind of answers sought (mostly just a number or a quantity), they are extremely open when it comes to choosing a framework within which they can be specified and dealt with. This means that most of the work involved in solving such a Fermi problem has to do with pre-mathematisation, with information and data collection, simplifications, idealisations and assumptions as the predominant components. Typically, there is a wide range of possible and feasible assumptions (What is the percentage of piano players in a city, and how often does a piano have to be tuned? What is the size of a girl's scalp, and how densely do hairs grow?). Correspondingly, there will be a very wide range of possible and reasonable answers. Let us take the hair problem as an example. Taking rough measures, we can estimate the size of a girl's scalp as the area of a hemisphere with radius 9 cm, so $2\pi 9^2$ cm^2, i.e., ca. 500 cm^2, with a range of reasonable areas between 350 cm^2 and 700 cm^2. The density of human hair depends on several factors including hair colour, gender, age, and geographical region, so let us imagine the question is posed in a suburban American grade 9 class. The number of hairs per cm^2 can be determined by counting the hairs in a circa 1 cm^2 region of some girls' scalps in this class. Assume that the average number turns out to be ca. 230 hairs per cm^2, with a range between 150 and 300. A rough estimate of the number of hairs is $500 \cdot 230 = 115{,}000$. If we calculate the extremes, we end up with a result between $350 \cdot 150 = 52{,}500$ and $700 \cdot 300 = 210{,}000$. The extremes differ by a factor of $2 \cdot 2 = 4$, but the order of magnitude is 100,000, so it is reasonable to say: The number of hairs of a girl (in this class) is (very roughly) 100,000.

2.5 Descriptive and prescriptive modelling

So far, we have outlined the fundamentals of mathematical modelling of contexts and situations that already exist within some extra-mathematical domain. The overarching purpose of this kind of modelling is to come to grips with aspects of the given context and situation that are of interest and significance to the modeller. Thus, the modeller may be interested in capturing and understanding the essential structure and organisation of the situation as well as the phenomena and behaviour it displays. The modeller may, furthermore, be interested in identifying the mechanisms that are (co-)responsible for the phenomena and the behaviour observed to explain these. The modeller may also want to predict future behaviour and the conditions under which a certain behaviour will or will not occur, perhaps in terms of different future scenarios. Sometimes, the ultimate goal of the modelling enterprise is to pave the way for and underpin the making of decisions based on an analysis of the situation offered by the model. It is common practice to use the term *descriptive modelling* for such modelling, aiming at capturing and coming to grips with *existing* contexts and situations.

There is, however, a different category of modelling purposes that do not primarily deal with capturing and understanding an existing reality but attempt to *create or organise reality*. This is the case when we want to design and construct material objects (e.g., roads, buildings, bridges, containers, utensils, tools, homeware, etc.; see example 4 about paper formats in Chapter 3) to fulfil certain requirements and specifications. It is also the case when we want to define financial instruments such as schemes for investment or for amortisation of loans, to construct rules of taxation (e.g., for income or consumption taxes), to exploit limited resources (e.g., materials or substances), or to harvest renewable resources (such as agricultural products, fish or seafood, wood from forests, or wind or water power) at favourable times. And the same holds when we want to determine a feasible or even optimal location of an object or a facility (e.g., a transmitter mast, a fire station or a hospital) within some domain or terrain or when we want to schedule activities, operations or processes to be conducted while allocating corresponding resources (e.g., vaccination programmes or production plans or routes and itineraries to be followed by transport vehicles). It is also the case when we want to design election procedures and ways to account for their results (such as distribution of seats in a parliament based on the voting outcomes) or when we want to define and introduce measures or concepts in practical or scientific contexts (e.g., speed and acceleration, density, kWh, pH, body mass indices, elasticity of variables in the economy, etc.). There is a plethora of examples of this kind of design problems.

In all such cases, mathematics is involved in providing prescriptions that help create or organise reality by structuring, changing or interfering with it. We follow Davis (1991) and Niss (2015) in using the term *prescriptive modelling* for such endeavours while observing that this type of modelling is sometimes also referred to as *normative modelling*.

The question arises whether the modelling cycle for descriptive modelling presented above also works in relation to prescriptive modelling. In prescriptive modelling, too, we take our point of departure in some extra-mathematical domain. The generating questions, however, are typically of a somewhat different nature from those in descriptive modelling as they focus on measures, decisions or actions that should or could be undertaken in order to create or organise the reality of the extra-mathematical domain that interests us. It is normally also necessary to undertake pre-mathematisation leading to a reduced situation accompanied by a specification of the design or decision questions we want to pose. Examples include: How can we introduce a relevant measure of "heaviness" for individuals? Where should the next general hospital in a given region be located? How can we design a container of milk which on the one hand fulfils certain requirements concerning its dimensions and on the other hand makes use of a minimum amount of material? What would be an appropriate measure of income inequality in a country? What should the life insurances premiums be for different out-payments in a certain sub-population of a country? And so on and so forth.

As is the case with descriptive modelling, the reduced situation *cum* questions must be mathematised into a mathematical situation *cum* questions belonging to some mathematical domain. Sometimes, the mathematised questions closely resemble those in descriptive modelling, and hence the same is true of the mathematical treatment. For instance, this is typically the case with optimisation questions and questions involving geometric design (e.g., the paper format design question (see example 4 in Chapter 3): We are to design a sequence of rectangular paper formats such that each sheet is produced by folding the previous sheet along a mid-point transversal and such that the ratio between the longer and shorter sides of any sheet is the same for all sheets, while requiring the largest sheet to have an area of $1m^2$).

However, in other situations, the mathematised question is trivial or very simple, thus requiring no or very little treatment. For example, this is the case if we have decided to measure the age profile of the inhabitants of a country by indicating the average age and the standard deviation and then want to determine the values of these indicators at a given date or for a given country in a given year. When it comes to de-mathematisation in prescriptive modelling, this varies greatly with the circumstances. Sometimes, de-mathematisation simply consists in attaching units to a numerical result, like when the optimal dimensions of a container of a certain shape must be specified by attaching units to the numbers arrived at. At other times, de-mathematisation involves much more interpretation, e.g., when making sense of the value of the Gini coefficient of economic inequality (Gini, 1912) in a given country and comparing it with the value of the coefficient in other countries.

To a large extent, the modelling cycle for descriptive modelling can be used to also capture prescriptive modelling, even though sometimes only parts of the cycle will be in play. The most important difference between descriptive and predictive modelling, however, normally lies in the evaluation of the resulting models (Niss, 2015). While it is possible in descriptive modelling to evaluate a model by validating its outputs, by assessing the structural properties of the model, by assessing its quantitative features and by comparing and contrasting it with other models representing the same context and situation, much of this may not be possible or may even not make sense when evaluating a model constructed for the purpose of prescriptive modelling. Validation of the model outputs by confronting them with existing reality will often be meaningless because the reality doesn't fully exist yet since the purpose of the modelling endeavour will typically be to design or structure reality. In a multitude of cases, it is simply not possible to evaluate a model constructed for prescriptive purposes by way of its outputs. Hence, it cannot be falsified. For example, one cannot validate BMI values and associated interval cut points by confrontation with reality. However, one or more of the other components of model evaluation – especially concerning its structural properties, quantitative features and comparison with other models – often are possible (Niss, 2015). Often, the consequences of prescriptive modelling (e.g., the tax revenue of a state according to a certain taxation model, the seat distribution of a parliamentary election based on a certain apportionment model, or the feasibility in practical terms of a certain location of a fire station) are open for public discussion with reference to personal or societal norms or values, which is also an evaluation of a different nature to the evaluation of models built for descriptive purposes. However, the bottom line is that the outcomes of prescriptive modelling enterprises are evaluated in terms of "fitness for purpose".

It is important to understand that descriptive and prescriptive modelling differ in the purposes pursued, not necessarily in the models produced. The very same model can occur as a result of descriptive as well as of prescriptive modelling, as is the case with the paper format and the amortisation of a loan situations (see examples 4 and 5 in Chapter 3). This is the reason why we speak of descriptive and prescriptive modell*ing*, respectively, not about descriptive or predictive models. Invoking an old distinction, we can say that the outcome of a descriptive modelling enterprise is a model *of* those aspects of the extra-mathematical domain that are captured by the model, whereas the outcome of a prescriptive modelling undertaking is a model *for* those aspects of the extra-mathematical world that have been created or shaped by the model and its implementation.

2.6 Cognitive aspects of modelling

So far, we have concentrated on presenting the basics of models and modelling in a general sense without having paid attention to cognition or education. However, for a person actually or potentially engaged in undertaking or learning modelling, there are, of course, several cognitive facets, and in some cases even barriers, at play in this endeavour. The presence of these facets depends on the overall setting for the modelling context and situation. First, it may happen that neither a context nor a situation is present from the beginning. If so, the person at issue only becomes a modeller once he or she has identified a context, situation and questions worth dealing with. In case a context and a situation are present from the beginning, how open is the situation? How much is given to the modeller at the outset, i.e., what are the boundary conditions for the undertaking? Is it up to the modeller her- or himself to pose the questions to be answered, to make assumptions about the situation, to seek and select relevant information, data, etc. pertaining to the situation, or is this already in place from the outset?

At any rate, the modeller first must make sense of and comprehend the extra-mathematical context and situation under consideration. This requires the modeller to construct a mental image of the situation and of the issues and questions to be considered, to form a view of the factors and components that are significant to the enterprise and the ones that aren't, and to mobilise whatever knowledge he or she possesses about the extra-mathematical domain pertaining to the situation. This part of the pre-mathematisation process, a clear mental image of the situation, is an indispensable first step in the modelling process. In other words, the modeller has to form a *situation model* (cf. Leiss et al., 2010, for a description of the origins and features of this construct). On this basis, the modeller moves on to specify the questions to which answers are being sought, to identify relevant variables, and to make appropriate assumptions about the situation and its elements in order to arrive at a reduced situation *cum* questions, possibly in the form of a real model (Blum, 1985). This part of the pre-mathematisation process involves the first instance of what is called *implemented anticipation* (Niss, 2010; Stillman & Brown, 2012, 2014; Niss, 2017; Czocher, 2018). This means that for the modeller to be able to make all the decisions and choices needed for preparing the situation for subsequent mathematisation, he or she must imagine which elements and relationships among them may, potentially, lend themselves to some form of mathematisation. In other words, the modeller must project her- or himself into a situation which doesn't quite exist yet as it awaits decisions, conclusions and steps further down the road. We will return to this notion in sections 2.8 and 6.2.

Once the reduced situation *cum* questions have been obtained, the modeller moves on to undertake perhaps the most crucial part of the entire modelling process: mathematisation. In this step, the modeller will have to decide which mathematical domain M and which objects and mathematical relationships would be suitable representatives of the extra-mathematical objects, features and relationships selected for inclusion in the reduced situation. The modeller will further have to specify the mathematical questions that would correspond to the extra-mathematical questions posed in D. All this involves the second and most significant instance of implemented anticipation by the modeller, who must project her- or himself into yet another situation which doesn't really exist yet. Already at this stage, the modeller will have to anticipate not only what mathematics might be suitable and available for capturing the essentials of the extra-mathematical situation but also how to use this mathematics to obtain conclusions that might eventually generate answers to the extra-mathematical questions that gave rise to the whole enterprise. This anticipation is strongly dependent on the range, nature and accessibility of the mathematical resources in the modeller's possession – not only those resources taken in a pure mathematical form but also their role in the modeller's previous experiences with models and modelling. A key issue here is that the mathematical resources a modeller can activate in a modelling situation must be thoroughly absorbed and connected in her or his mind. It is difficult – albeit not entirely impossible – to bring mathematical concepts, theories and methods that do not (yet) belong to one's resources to bear on the mathematisation step before this has been taken. In other words, we are faced with a bit of a paradox: "In order to undertake mathematisation, you already have to be able to undertake mathematisation!"

In a paper by Treilibs et al. (1980), they considered what they called *model formulation* consisting of (quoted from Burkhardt, 2018, p. 63; our italics):

> [G]*enerating variables* – the ability to generate the variables or factors that might be pertinent to the problem situation; *selecting variables* – the ability to distinguish the relative importance of variables in the building of a good model; *specifying questions* – the ability to identify the specific questions crucial to the, typically ill-defined realistic problem; *generating relationships* – the ability to identify relationships between the variables inherent in the problem situation; *selecting relationships* – the ability to distinguish the applicability of possible relationships to the problem situation.

It is interesting to notice that this a specification of what it takes for someone to be able to carry out an amalgamation of the pre-mathematisation and mathematisation phases of the modelling process.

The mathematisation step is complete when the mathematical objects, properties, relationships and questions meant to represent the (possibly reduced) extra-mathematical situation *cum* questions have been selected. Then the modeller will seek answers to the mathematical questions that have arisen during the mathematisation step. This means that the modeller will have to engage in mathematical treatment of the relevant entities in M, especially in mathematical problem solving. As the mathematical problems corresponding to the questions posed may be wildly varying in complexity and technical sophistication, ranging from trivial, routine problems to deep and challenging problems that nobody has solved before, it goes without saying that there is a similar range of variation in the cognitive demands on the modeller when it comes to mathematical treatment. The literature on mathematical problem solving has dealt with these demands for three quarters of a century. This is not the place to outline, let alone detail, the nature of these demands. Suffice it to mention that the demands concern at least the following spectrum of components (Schoenfeld, 1992): the knowledge base, problem solving strategies, monitoring and control, beliefs and affects, and practices. Here, the ability to devise and implement a problem-solving strategy seems to be the core component, the elements of which are: familiarising oneself with the original problem situation, analysing it, exploring possible solution paths, settling on a plan, implementing the plan and verifying the outcomes. Here we encounter the third instance of implemented anticipation that the modeller needs to produce. When faced with a well-defined non-trivial mathematical problem, the problem solver has to imagine, prior to the fact, possible approaches to attack the problem while anticipating the ways in which a given approach might eventually be conducive to obtaining a well-justified full or partial solution to the problem and, hence, answers to the mathematical questions posed within M.

Once the modeller has obtained mathematical answers to the mathematical questions resulting from the mathematisation step, the modeller must de-mathematise the answers, i.e., translate them back and interpret them in terms of answers to the extra-mathematical questions pertaining to the reduced situation. Sometimes de-mathematising the mathematical answers is straightforward and simple; sometimes it is not simple at all, typically if substantial interpretation is required to make sense of the outcomes within the extra-mathematical domain. Take, for example, a probabilistic weather forecasting model that concludes that the probability of rain in the capital tomorrow afternoon is 60%. How can that statement be interpreted in real-life terms, and what consequences should it have for the family's picnic plans? Will it rain everywhere 60% of the time? Is the probability of it raining at a specific point in the capital 60%?

Having de-mathematised the mathematical outcomes, the modeller will need to validate the answers obtained vis-à-vis the extra-mathematical questions that gave rise to the modelling enterprise in the first place, as described in section 2.2. In case the answers are discarded as unsatisfactory, the modeller is faced with the necessity of improving the model by way of some form of modification or amendment or by building an entirely new model by taking a different approach. In either case, the validation of the answers lead to a negative evaluation of the model itself. As outlined in section 2.2, there are other ways to evaluate a model than by validating the answers. If the evaluation of the model leads to it being discarded, the modeller is faced with the challenge of coming up with a new model. Taking this path gives rise to a fourth – and strong – instance of implemented anticipation because the modeller has to imagine how this might be done in a manner that avoids the problems responsible for the failure of the first model and to anticipate and implement the ideas thus generated. The Blum-Leiss version of the modelling cycle shown in Figure 2.8 illustrates, in an ideal-typical way, these steps from a cognitive point of view.

Once again, it cannot be stressed enough that this depiction of the cognitive processes involved in performing modelling is an analytic reconstruction of what must happen in principle. It is not at all a description of the path a concrete modeller will necessarily take in actual practice.

2.7 Modelling in mathematics education – a brief historical sketch

Thus far, we have been dealing with the theoretical and cognitive aspects of mathematical modelling, whereas nothing has been said about mathematical modelling in the context of mathematics education. The time has now come to address this topic, with an emphasis on the actual or potential place and role of modelling in mathematics curricula, teaching and learning as well as in mathematics education research.

We know that already ancient Egypt and Mesopotamia, for educational reasons, took an interest in what we usually call *word problems*, i.e., problems in which some questions are asked as part of a short, verbally formulated narrative about a more or less idealised real-world situation, accompanied by a few pieces of quantitative information. The questions typically require activation of arithmetic, algebraic or algebraic-geometric solving processes. As an example, consider problem 72 of the Egyptian Ahmes papyrus, dated between 2000 and 1800 BC: "How many loaves of strength 45 are equivalent to 100 loaves of strength 10?" where strength of a loaf is the reciprocal of grain density. Word problems later became popular in medieval China, India, in the Muslim world and in Europe. As proposed by Verschaffel et al. (2000), it may be natural to consider word problems as a special type of modelling tasks – a rather stylised one – even if the modelling terminology was not used before the last decades of the 20th century. If we accept word problems as a special form of modelling tasks – but many protagonists in the didactics of mathematical modelling do not – mathematics education has always had some kind of modelling on the agenda of teaching. More will be said about word problems in section 2.9.

Traditionally, however, the term "modelling" has been reserved for much less stylised situations in which the properties, features and attributes of the extra-mathematical contexts and situations at issue play a significant role in the modelling enterprise, in contradistinction to what is typically the case with word problems. A key figure in pleading for the serious inclusion of all aspects of the mathematical modelling process in mathematics education, Henry Pollak, who is an early protagonist in mathematical applications, models and modelling (Pollak, 1968, 1969, 1979), has investigated (Pollak, 2003) the roots of mathematical modelling (or the equivalent notion of model building) – not only of mathematical models – in mathematics teaching and learning. Modelling was introduced in the USA at both college and school levels in the mid-1960s by the Committee on the Undergraduate Program in Mathematics (1966) in *A Curriculum in Applied Mathematics: Report of Ad Hoc Committee on Applied Mathematics* and by the School Mathematics Study Group (SMSG) Committee on Mathematical Models (1966) in *Report of the Modeling Committee*, respectively (see also Burkhardt & Pollak, 2006).

The 1970s saw an upsurge of mathematical modelling in experimental teaching programmes, primarily at the tertiary level, in different parts of the world, including Australia, Denmark, Germany, the Netherlands and the UK, many of which were presented in reports and journal papers. At the Third International Congress on Mathematical Education (ICME 3), held in Karlsruhe, Germany, in 1976, Henry Pollak was in charge of the section on the relationship between mathematics and other school subjects in which the importance of mathematical modelling was explicitly stressed (Pollak, 1979). From then on, all the ICMEs contained significant programme elements focusing on mathematical applications and modelling. In the UK, the Shell Centre for Mathematical Education was established at Nottingham University in 1967; when Hugh Burkhardt became its director in 1976, a focus on mathematical modelling was adopted and developed further (for more details, see Burkhardt, 2018; for materials, see section 5.6).

In the USA, a non-profit organisation, Consortium for Mathematics and Its Applications (COMAP, www.comap.com), was set up by Solomon Garfunkel in

1980 to further mathematical modelling in mathematics education. Since then, COMAP has been instrumental in designing and disseminating a variety of curriculum and teaching materials for schools and colleges in the US (see sections 5.6 and 7.4).

Mathematical modelling gradually obtained an increasingly strong position within the growing international community, cultivating an interest in mathematical applications and modelling. This community established a series of International Conferences on the Teaching of Mathematical Modelling and Applications (the ICTMAs), the first of which was held in Exeter, UK, in 1983. The most recent one, ICTMA 19, was held in Hong Kong in 2019. In 1999, the community around the conferences decided to develop an organisational framework for its activities, choosing the name The International Community of Teachers of Mathematical Modelling and Applications, generating also the acronym ICTMA. In 2003, ICTMA was accepted as an Affiliated Study Group of the International Commission on Mathematical Instruction (ICMI).

In the first wave of work on mathematical applications and modelling, which we might roughly date to the years 1970–1995 (also see the survey by Blum & Niss, 1991; Niss, 2018), the primary foci were: (1) making a plea and giving arguments for the inclusion of these components in mathematics teaching and learning at all levels, while also analysing, on theoretical grounds, how this may take place, as well as establishing a terminology for dealing with applications and modelling issues; (2) presenting applications and modelling cases reporting on how people had implemented applications and modelling activities in actual curricula and in actual teaching to provide an existence proof of the viability of these ideas in real life; and (3) collecting and publishing applications and modelling examples, which interested school or college teachers might either use directly or take inspiration from in their own teaching. One might describe this first wave as one in which the theoretical foundation thus laid for applications and modelling constituted the (theoretical) research part of the work, while the practical, experimental and empirical activities gave flesh and blood to the theoretical consideration without, at that time, involving empirical research proper to a substantive extent.

The second wave – by and large from 1995 onwards – is characterised (Niss, 2018) by three features: (4) modelling became the predominant focus point, whereas applications not directly involving the modelling process lost significance; (5) the very process and sub-processes of modelling as well as the corresponding modelling competencies (see Chapter 4) attracted substantive attention, as did the barriers encountered by students when undertaking modelling activities (see Chapters 5 and 6); and (6) empirical research on the teaching and learning of modelling gained massive prominence among those working in the field and that this research was acknowledged also by the mathematics education community at large. We can say that now, in 2019–2020, we are still in this phase. While ICMI Study 14 on Modelling and Applications in Mathematics Education (Blum et al., 2007) gave a summary of the state of the art in the field until the beginning of our century, we will, in this book, also take stock of more recent achievements and developments.

2.8 Mathematical modelling in mathematics teaching and learning – why, what and how

There are basically two different overarching reasons for including mathematical modelling as a significant component of mathematics teaching and learning at all education levels.

The first reason is to do with the fact that mathematics plays an immensely important role in understanding and dealing with the world around us. As this role is brought about by way of mathematical modelling, enabling people to put the mathematics they have learnt and understood to use in extra-mathematical contexts and situations – i.e., undertaking mathematical modelling – should be a goal in and of itself of mathematics education. In that way, mathematics ought to help students to come to grips with the world in which we live and to better master real-world situations stemming from everyday life as well as from other school subjects or from their future professions or fields of study. It should also be a goal to educate students to see and understand the use and misuse of mathematics in society. All this requires the development of a general mathematical modelling competency (see chapter 4). In simplistic terms, we might label this reason *mathematics for the sake of modelling*, even though modelling is usually not the only goal of mathematics education.

The second reason is to employ modelling as a means for something else, above all for supporting the learning of mathematics, by offering motivation for its study as well as interpretation, meaning, proper understanding and sustainable retention of its concepts, results, methods and theories and at the same time furthering significant mathematical competencies such as problem solving and reasoning (see section 4.5). In that way, modelling also contributes to generating an adequate and balanced image of mathematics as a discipline, from both historical and contemporary perspectives. Again, in simplistic terms we might label this reason *modelling for the sake of mathematics* (learning and appreciation).

These two reasons are in no way contradictory to one another – both can be pursued in the same actualisation of teaching and learning of mathematics. They are, however, analytically distinct, and they do give rise to different consequences in terms of priorities and activities in the design and orchestration of mathematics teaching.

If the former reason is being invoked, according to which it is an obligation and a goal for mathematics teaching and learning that students become able to undertake descriptive and prescriptive mathematical modelling in a variety of contexts and settings, the full modelling cycle, in one version or another, whether explicit or implicit, is necessarily placed on the agenda of mathematics education. This implies, among other things, that numerous aspects of the extra-mathematical domains under consideration must be considered as part of the modelling process. It is not sufficient to concentrate on the intra-mathematical elements of the models, including the pure mathematical problem solving involved in dealing with them. In other words, students will have to be engaged in working on the entire modelling cycle. As Blomhøj and Jensen (2003) have pointed out, this can take place in two different ways: *holistically* or *atomistically*. In the holistic approach, students are working on modelling tasks in a manner that involves all the components of the modelling cycle at the same time. In the atomistic approach, students are working on modelling tasks in a way that zooms in on only one or a few such components at a time, typically different components in different tasks. In Chapters 5 and 6, the didactic implications of these approaches are discussed.

In case the latter reason is being invoked, according to which the primary role of modelling is to support the teaching and learning of mathematics as a discipline, the choice of extra-mathematical domains to be subjected to modelling is subsumed under the main purpose of providing motivation, illustration, interpretation and meaning to the mathematical entities and processes placed on the agenda of the teaching in the given context. Similarly, the role assigned to the modelling cycle and its different components in the context is determined by the needs that are supposed to be served by involving modelling in the teaching and learning of mathematics. For motivational and other learning purposes, it might even be legitimate to use dressed-up so-called "word problems" (see section 2.9) where modelling essentially consists in undressing the problem so as to make the mathematical core visible and, once a solution to the mathematical problem has been obtained, to interpret it within the situation given. However, for developing an adequate and balanced image of mathematics, it is essential to also, from time to time, address authentic modelling problems by running through the whole modelling cycle.

One such way of placing modelling in the service of the teaching and learning of mathematics involves what Lesh and Doerr (2003) call *model-eliciting activities* (MEAs). In such activities, students are placed in extra-mathematical situations in which they are to capture essential features of the situation by (re-)inventing mathematical concepts such as ratio, rate of change, slope, scaling, average, standard deviation and so on so as to underpin the formation of these mathematical concepts (also see section 6.9). This is similar to Freudenthal's idea of *guided re-invention* (Freudenthal, 1991). A related notion is that of *emergent modelling* introduced by Gravemeijer (2007). By attempting to come to grips with a given extra-mathematical situation by inventing or using mathematical descriptions of it – i.e., a model *of* the situation – students gradually accumulate a repertoire of such descriptions which eventually can be structured and directly employed to dealing with similar new situations so that students now possess a model *for* a large class of such situations. The latter aspect can also serve the purpose of building a capacity in students for undertaking implemented anticipation with particular regard to mathematisation.

The various aims pursued by way of mathematical modelling activities have given rise to the distinction between various *perspectives* of modelling, as Kaiser et al. (2006) have called it. According to the purposes of modelling they speak of "applied", "educational", "socio-critical", "epistemological", "pedagogical" or "conceptual" modelling. Blum (2015) suggests regarding perspectives as a triple, consisting of an *aim* for modelling, suitable modelling *examples* and a *modelling cycle* suitable for visualisation of the aims. He also distinguishes (pp. 82–83) between

30 Conceptual and theoretical framework

different aspects of sense-making via modelling, corresponding to the various perspectives. In a similar approach, Abassian et al. (2019) characterise five important perspectives by goals, model definition, modelling cycle and task design.

2.9 Word problems

Since word problems play a special and somewhat controversial part in the discussion, research and practice of mathematical modelling, it is warranted to say a few words about this notion. The term "word problem" is widely used in mathematics, but clear definitions are scarce. Wikipedia offers the following definition:

> In science education, a *word problem* is a mathematical exercise where significant background information on the problem is presented as text rather than in mathematical notation. As word problems often involve a narrative of some sort, they are also referred to as *story problems* and may vary in the amount of language used.
> *(https://en.wikipedia.org/wiki/word_problem(mathematics_education),*
> *accessed 28 October 2019)*

This definition focuses entirely on the use of words instead of mathematical notation. This would also make "Show that every metric space is a Hausdorff space" a word problem, even though this clearly does not fall under this category according to the common understanding of "word problem" in mathematics education. To capture that understanding, we propose the following definition:

A *word problem* poses – typically within an educational or recreational setting – a question concerning a real, idealised or imagined extra-mathematical context and situation, the answering of which requires some sort and degree of mathematical problem solving.

While the answer to a word problem is supposed to be unique, the mathematical treatment leading to this answer may – in principle, although not necessarily in practice – vary considerably. The problem, including its background information, is presented by way of a few lines of text in ordinary language, supplemented by elementary numbers and units. The background information presented is both necessary and sufficient for the solution of the problem. The primary purpose of a word problem is not to understand the given context by means of mathematics but to present the underlying mathematical problem in an interesting and motivating wrapping.

Word problems differ in the significance of the extra-mathematical context and situation for the problem posed and for the solution process undertaken. At one end of the spectrum, the extra-mathematical context has no real significance because its role is to dress up – disguise – an intra-mathematical problem that constitutes the kernel of the task (Pollak, 1979, calls such problems "whimsical"). Accordingly, the solution consists of undressing the problem to reveal its intra-mathematical structure and substance and then solving it. At the other end of the spectrum, the extra-mathematical context and situation are indeed significant and cannot be discarded, but the problem solver is not supposed to find any information or data him- or herself or make any additional assumptions, idealisations or simplifications beyond those stated in the problem description. In other words, no pre-mathematisation needs to be undertaken prior to the mathematisation, which is typically stylised and uniquely determined. Similarly, answer validation and model evaluation only concern the relevance and correctness of the mathematisation and the mathematical treatment performed. Furthermore, word problems differ with respect to the realism of the extra-mathematical context. Is the context embedded in a real domain; is the problem authentic, or does it refer to some highly idealised or fictitious domain, with correspondingly idealised or fictitious questions?

Word problems of the latter category have been widely ridiculed in various kinds of literature. The most famous example is found in a letter which the French writer Gustave Flaubert wrote to his sister Caroline in 1841, who was studying trigonometry at the time (Flaubert, 1887):

> A ship sails the ocean. It left Boston with a cargo of wool. It grosses 200 tons. It is bound to Le Havre. The mainmast is broken, the cabin boy is on deck, there are 12 passengers aboard, the wind is blowing East-North-East, the clock points to a quarter past three in the afternoon. It is the month of May. How old is the captain?

This, of course, is an extreme example, but less extreme examples abound. This may be one of the reasons why research has shown that many students tend to treat most word problems as if they are dressed-up problems and rush off to manipulate the numbers given in the problem statement by mathematical means, while entirely suspending any kind of sense-making related to the extra-mathematical context and situation (Verschaffel et al., 2000; Verschaffel et al., 2010; Jankvist & Niss, 2019. See section 5.2 for more details).

As one means of "suspending the suspension of sense-making", various researchers, especially Greer (1993, 1997), Greer et al. (2007), Verschaffel et al. (2010) and Verschaffel et al. (2000), have made a plea for perceiving word problems as a certain, albeit restricted, kind of modelling problems. In contrast to many other researchers in the area of the didactics of mathematical modelling, we endorse this perspective but nevertheless maintain the necessity of distinguishing clearly between word problems and full-fledged modelling problems. The processes involved in solving word problems are: (1) reading and understanding the problem presentation; (2) identifying the mathematical problem embedded in the problem presentation; (3) solving the mathematical problem, primarily by performing the mathematical operations entailed by the problem; and (4) presenting the solution and writing and justifying the answer. In contrast, a full-fledged modelling process involves all the phases of the modelling cycle, including pre-mathematisation with its multitude of aspects, mathematisation, mathematical treatment, de-mathematisation, validation of model outcomes and evaluation of the entire model.

Notes

1 The word "mapping" is in inverted commas because f is usually not a mapping in a strict mathematical sense since it is not the case that every single object in D is mapped onto one (and only one) object in M. In other words, "mapping" is used as an analogy.
2 As is commonplace in mathematics education, we have chosen to use the term "linear" for such functions, even though they are not really linear but affine in the sense of linear algebra.

References

Abassian, A., Safi, F., Bush, S. & Bostic, J. (2019). Five different perspectives on mathematical modeling in mathematics education. In: *Investigations in Mathematics Learning*. DOI: 10.1080/19477503.2019.1595360

Blomhøj, M. & Jensen, T.H. (2003). Developing mathematical modelling competence: Conceptual clarification and educational planning. In: *Teaching Mathematics and Its Applications* 22(3), 123–139.

Blomhøj, M. & Jensen, T.H. (2007). What's all the fuss about competencies? In: W. Blum, P.L. Galbraith, H.-W. Henn & M. Niss (Eds.), *Modelling and Applications in Mathematics Education* (pp. 45–56). New York, NY: Springer.

Blum, W. (1985). Anwendungsorientierter Mathematikunterricht in der didaktischen Diskussion. In: *Mathematische Semesterberichte* 32(2), 195–232.

Blum, W. (2015). Quality teaching of mathematical modelling: What do we know, what can we do? In: S.J. Cho (Ed.), *The Proceedings of the 12th International Congress on Mathematical Education: Intellectual and Attitudinal Challenges* (pp. 73–96). New York, NY: Springer.

Blum, W., Galbraith, P.L., Henn, H.-W. & Niss, M. (Eds.) (2007). *Modelling and Applications in Mathematics Education: The 14th ICMI Study*. New York, NY: Springer.

Blum, W. & Leiss, D. (2007). How do students and teachers deal with mathematical modelling problems? The example "filling up". In: C.R. Haines, P.L. Galbraith, W. Blum & S. Khan (Eds.), *Mathematical Modelling (ICTMA 12): Education, Engineering and Economics* (pp. 222–231). Chichester: Horwood Publishing.

Blum, W. & Niss, M. (1991). Applied mathematical problem solving, modelling, applications, and links to other subjects: State trends and issues in mathematics instruction. In: *Educational Studies in Mathematics* 22(1), 37–68.

Borromeo Ferri, R. (2006). Theoretical and empirical differentiations of phases in the modelling process. In: *ZDM: The International Journal on Mathematics Education* 38(2), 86–95.

Burkhardt, H. (2018). Ways to teach modelling: A 50 year study. In: *ZDM: The International Journal on Mathematics Education* 50(1 + 2), 61–75.

Burkhardt, H. & Pollak, H. (2006). Modelling in mathematics classrooms: Reflections on past developments and the future. In: *ZDM: The International Journal on Mathematics Education* 38, 178–195.

Committee on the Undergraduate Program in Mathematics (1966). *A Curriculum in Applied Mathematics: Report of Ad Hoc Committee on Applied Mathematics*. Washington, DC: Mathematical Association of America.

Czocher, J. (2018). How does validating activity contribute to the modelling process? In: *Educational Studies in Mathematics* 99(2), 137–159.

Davis, P.J. (1991). Applied mathematics as a social instrument. In: M. Niss, W. Blum & I. Huntley (Eds.), *Teaching Mathematical Modelling and Applications* (pp. 1–9). Chichester: Ellis Horwood.

Flaubert, G. (1887). *Correspondance, première série (1830–1850), 16 mai 1841*. Paris: G. Charpentier et Cie, Éditeurs.

Freudenthal, H. (1991). *Revisiting Mathematics Education: China Lectures*. Dordrecht: Reidel Publishers.

Galbraith, P. & Stillman, G. (2006). A framework for identifying student blockages during transitions in the modelling process. In: *ZDM: The International Journal on Mathematics Education* **38**(2), 143–162.

Gini, C. (1912). *Variabilità e Mutabilità. Contributo allo Studio delle Distribuzione e delle Relazioni Statistiche*. Bologna: C. Cuppini.

Gravemeijer, K. (2007). Emergent modelling as a precursor to mathematical modelling. In: W. Blum, P.L. Galbraith, H.-W. Henn & M. Niss (Eds.), *Modelling and Applications in Mathematics Education: The 14th ICMI Study* (pp. 137–144). New York, NY: Springer.

Greer, B. (1993). The mathematical modelling perspective on wor(l)d problems. In: *Journal of Mathematical Behavior* **12**, 239–250.

Greer, B. (1997). Modelling reality in mathematics classrooms: The case of word problems. In: *Learning and Instruction* **7**(4), 293–307.

Greer, B., Verschaffel, L. & Mukhopadhay, S. (2007). Modelling for life: Mathematics and childrens' experience. In: W. Blum, P.L. Galbraith, W. Henn & M. Niss (Eds.), *Modelling and Applications in Mathematics Education: The 14th ICMI Study* (pp. 89–95). New York, NY: Springer.

Jankvist, U.T. & Niss, M. (2019). Upper secondary students' difficulties with mathematical modelling. In: *International Journal of Mathematical Education in Science and Technology*. DOI: 10.1080/0020739X.2019.1587530

Kaiser, G., Blomhøj, M. & Sriraman, B. (2006). Towards a didactical theory for mathematical modelling. In: *ZDM: The International Journal on Mathematics Education* **38**(2), 82–85.

Kaiser, G. & Stender, P. (2013). Complex modelling problems in co-operative, self-directed learning environments. In: G. Stillman, G. Kaiser, W. Blum & J. Brown (Eds.), *Teaching Mathematical Modelling: Connecting to Research and Practice* (pp. 277–293). Dordrecht: Springer.

Leiss, D., Schukajlow, S., Blum, W., Messner, R. & Pekrun, R. (2010). The role of the situation model in mathematical modelling: Task analyses, student competencies, and teacher interventions. In: *Journal für Mathematik-Didaktik* **31**, 119–141.

Lesh, R. & Doerr, H. (2003). *Beyond Constructivism: Models and Modeling Perspectives on Mathematics Teaching, Learning and Problem Solving*. Mahwah, NJ: Lawrence Erlbaum.

Niss, M. (1989). Aims and scope of applications and modelling in mathematics curricula. In: W. Blum, J.S. Berry, R. Biehler, I.D. Huntley, G. Kaiser-Messmer & L. Profke (Eds.), *Applications and Modelling in Learning and Teaching Mathematics* (pp. 22–31). Chichester, UK: Ellis Horwood.

Niss, M. (2010). Modeling a crucial aspect of students' mathematical modeling. In: R. Lesh, P.L. Galbraith, C.R. Hainses & A. Hurford (Eds.), *Modeling Students' Mathematical Modeling Competencies: ICTMA 13* (pp. 43–59). New York, NY: Springer.

Niss, M. (2015). Prescriptive modelling: Challenges and opportunities. In: G. Stillman, W. Blum & M.S. Biembengut (Eds.), *Mathematical Modelling in Education, Research and Practice: Cultural, Social and Cognitive Influences* (pp. 67–79). Cham, Heidelberg, New York, Dordrecht, London: Springer.

Niss, Ma. (2017). Obstacles related to structuring for mathematization encountered by students when solving physics problems. In: *International Journal of Science and Mathematics Education* **15**, 1441–1462.

Niss, M. (2018). Advances in research and development concerning mathematical modelling in mathematics education. In: F.-J. Hsieh (Ed.), *Proceedings of the 8th ICMI East Asia Regional Conference on Mathematics Education: ICMI-EARCOME8, Taipei, Taiwan, May 7–11* (Vol. 1, pp. 26–36). Taipei: National University of Taiwan.

Perrenet, J. & Zwanefeld, B. (2012). The many faces of the modeling cycle. In: *Journal of Mathematical Modelling and Applications* **1**(6), 3–21.

Pollak, H. (1968). On some of the problems of teaching applications of mathematics. In: *Educational Studies in Mathematics* **1**(1–2), 24–30.

Pollak, H. (1969). How can we teach applications of mathematics? In: *Educational Studies in Mathematics* **2**, 393–404.

Pollak, H. (1979). The interaction between mathematics and other school subjects. In: UNESCO (Ed.), *New Trends in Mathematics Teaching IV* (pp. 232–248). Paris: UNESCO.

Pollak, H. (2003). A history of the teaching of modeling. In: G.M.A. Stanic & J. Kilpatrick (Eds.), *A History of School Mathematics* (Vol. 1, pp. 647–671). Reston, VA: National Council of Teachers of Mathematics.

Schoenfeld, A. (1992). Learning from Instruction. In: D.A. Grouws (Ed.), *Handbook of Research on Mathematics Teaching and Learning* (pp. 334–370). New York, NY: Macmillan Publishing Company.

School Mathematics Study Group: Committee on Mathematical Models (1966). Report of the modeling committee. In: *Tentative Outlines of a Mathematics Curriculum for Grades 7, 8, and 9. SMSG Working Paper*. Stanford, CA: SMSG.

Stillman, G. & Brown, J. (2012). Empirical evidence for Niss' implemented anticipation in mathematising realistic situations. In: J. Dinyal, L.P. Cheng, S.F. Ng (Eds.), *Mathematics Education: Expanding Horizons* (Vol. 2, pp. 682–689). Adelaide: MERGA.

Stillman, G. & Brown, J. (2014). Evidence of implemented anticipation in mathematising by beginning modellers. In: *Mathematics Education Research Journal* **26**, 763–789.

Treilibs, V., Burkhardt, H. & Low, B. (1980). *Formulation Processes in Mathematical Modelling*. Nottingham: Shell Centre Publications.

Verschaffel, L., Greer, B. & De Corte, E. (2000). *Making Sense of Word Problems*. Lisse: Swets & Zeitlinger.

Verschaffel, L., Van Dooren, W., Greer, B. & Mukhopadyay, S. (2010). Reconceptualising word problems as exercises in mathematical modelling. In: *Journal für Mathematik-Didaktik* **31**, 9–29.

3
MODELLING EXAMPLES

Introduction

In order to provide flesh and blood to the general and somewhat abstract presentation and discussion of models and modelling offered in Chapter 2, this chapter is devoted to a detailed exposition of a collection of modelling examples, all of which are both accessible at a mathematical level within the range of secondary school curricula in most parts of the world and manageable insofar as they can be discussed together with students within one or a few lessons. The examples in this chapter cover a range of real-world situations and mathematical content, as well as all educational levels from primary to upper secondary school. The following examples are ordered according to their cognitive and technical complexity.

Example 1: Uwe Seeler's foot

The problem situation

Uwe Seeler (born in 1936) is a famous German soccer player. He played 72 times internationally and participated in 4 world championships. In 2005, the city of Hamburg, where Seeler was born and where he played during his whole carrier, in his honour erected a big, bronze sculpture in front of the soccer stadium, showing his right foot (see Figure 3.1, taken from Vorhölter et al., 2019). The sculpture is, according to the city of Hamburg (see www.kulturkarte.de/hamburg/16015unsuwe), 5.15 m long, 2.30 m wide and 5.30 m high.

How big would a full statue of Uwe Seeler be if this sculpture was to be his right foot? In reality, Seeler's height is 1.68 m, and his (European) shoe size is 42.

36 Modelling examples

FIGURE 3.1 The Uwe Seeler foot statue (made by the artist Brigitte Schmittges) in Hamburg

Solving the problem

The first approach: A shoe size of 42 means approximately a 26.5-cm foot length. Since the sculpture is 5.15 m long, it is approximately 19.5 times as long as his real foot. Therefore, the scale is approximately 19.5:1, and hence a statue of Uwe Seeler with this sculpture as his foot would be gigantic, namely about 19.5 · 1.68 m ≈ 32.8 m tall.

The second approach: For Uwe Seeler, the ratio of his height and his shoe size is 4. This is actually the average for men in general. The European shoe size is approximately 1.5 times the shoe length, so the height of a man is approximately 6 times his shoe length. The shoe length is roughly 5% larger than the foot length; hence a foot length of 5.15 m would mean a shoe length of approximately 5.41 m. Therefore, the giant Uwe Seeler statue would be approximately 6 · 5.41 m ≈ 32.5 m tall.

Validating the solutions

How precise are our results? In the first approach, we used the information that a size 42 shoe means a foot length of 26.5 cm. This is a rough measure, and we do not know Seeler's actual foot length. If we are more cautious, we can say that Seeler's foot length will most likely be between 26 cm and 27 cm. The measures of the statue are all rounded to the nearest 5 cm, so the actual length will be between 5.125 m and 5.175 m. This means that the scaling factor will be between 5.175:26 ≈ 19.9 and 5.125:27 ≈ 18.9. Uwe Seeler's height is rounded to the nearest cm. Therefore, the resulting height of the giant statue lies in the interval between 18.9 · 1.675 m ≈ 31.6 m and 19.9 · 1.685 m ≈ 33.5 m. Similar considerations apply to the second approach. A reasonable answer will be: A full statue of Uwe Seeler using the given ratio of his foot would be approximately between 31.5 m and 33.5 m tall – or more cautiously between 31 m and 34 m.

On the internet, we can find the information that the artist wanted to create a sculpture with a scale of 20:1. That would mean the full Seeler statue would be 20 times Seeler's real height, that is, 33.6 m. However, this result is probably not precise because both the scaling factor of 20 and Seeler's height are approximate. Taking into account that Seeler was nearly 70 when the sculpture was made, his actual height is likely to have decreased a bit (people become shorter with age after 50), so his giant statue would have been shorter than if made at the time when he was an active soccer player.

Concluding remarks

This example shows how simple scaling can be used in elementary modelling and that scaling with different points of departure can be used to validate the model outcomes and eventually to evaluate the underlying model. It further shows how uncertainties in the initial measurements give rise to uncertainty estimates of the final results. When modelling, one of the principles of numerical analysis is important to keep in mind: The final result can never be more precise than the data used in the solution approach, and all uncertainties in the data increase the uncertainty of the result.

Reference

Vorhölter, K., Krüger, A. & Wendt, L. (2019). Metacognition in mathematical modeling: An overview. In: S.A. Chamberlin & B. Sriraman (Eds.), *Affect in Mathematical Modeling* (pp. 29–51). Cham: Springer.

Example 2: Filling up

The problem and a first solution

We imagine the following situation: Someone, we call her Mrs. Stein, must fill up her car. Where should she do that? We reduce the complexity of this situation by taking only two petrol stations into account. Mrs. Stein knows that the petrol in a station some distance away is much cheaper than in the station around the corner. Is it worthwhile to drive to the distant station? This is a kind of problem situation that can be encountered in other contexts as well – for instance, if someone has to decide whether it is worthwhile to drive to a supermarket outside of town because some goods are cheaper there than in the nearby market. We are going to analyse the filling up problem in considerable detail (see Blum & Leiß, 2005 for an early version of this problem, based on an authentic case).

38 Modelling examples

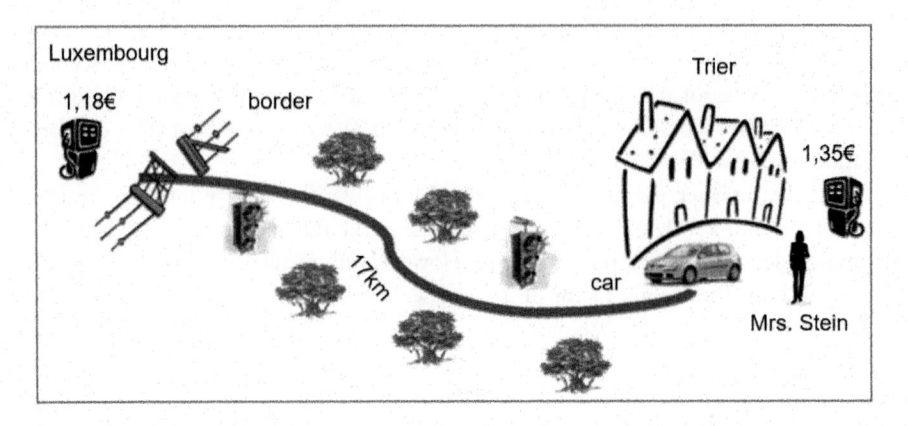

FIGURE 3.2 Map of Mrs. Stein's route

Assume that Mrs. Stein lives in Trier (Germany) where the petrol in the nearby station costs 1.35 € per litre. The petrol in the station just behind the border of Luxembourg, 17 km from where Mrs. Stein lives, costs only 1.18 € per litre (see Figure 3.2).

What we must know to be able to compare the prices for filling up in Trier or in Luxembourg are two parameters from Mrs. Stein's car: the consumption rate and the tank volume. Assume that the car consumes 0.07 litres per km on such a trip. The drive to the station in Luxembourg (34 km for both ways) costs:

$$34 \text{ km} \cdot (0.07 \text{ l/km}) \cdot 1.18 \text{ €/l} \approx 2.80 \text{ €},$$

supposing the petrol in the tank was already from that station (otherwise, we would have to calculate the costs for driving there and back separately, and the drive would cost $17 \cdot 0.07 \cdot 0.17 \text{ €} \approx 0.20 \text{ €}$ more). If we assume a tank volume of 45 litres and that Mrs. Stein arrives at the station with a nearly empty tank (creating a certain risk, especially when driving through the forest to Luxembourg), then filling up in Trier costs:

$$45 \text{ l} \cdot 1.35 \text{ €/l} = 60.75 \text{ €}$$

(if the station is just around the corner, or if Mrs. Stein passes by the station anyway), whereas filling up in Luxembourg costs:

$$45 \text{ l} \cdot 1.18 \text{ €/l} = 53.10 \text{ €},$$

implying that the total cost for filling up in Luxembourg is:

$$53.10 \text{ €} + 2.80 \text{ €} = 55.90 \text{ €}.$$

This is 4.85 € cheaper than filling up in Trier, so the answer is: If money is the only issue, it is worthwhile to drive to Luxembourg.

An evaluation of the first solution and moving to a second solution

Let us look at the problem and its solution a bit more carefully. Filling up in Luxembourg not only costs money for driving forth and back; in addition, the tank after the trip contains only 45 l minus the volume consumed on the 17 km return trip, which is 17 km · (0.07 l/km) = 1.19 l. Moreover, these 1.19 l had to be in the tank, as a minimum, before driving to Luxembourg. So a direct comparison with filling up in Trier must take into account that after the trip only 45 l − 2.38 l = 42.62 l more petrol is in the tank than before the trip. The corresponding cost for filling up this volume in Trier is only:

42.62 l · 1.35 €/l = 57.54 €.

This must be compared with the cost of filling up in Luxembourg, that is, 53.10 €. This is still cheaper than in Trier, but a fair comparison means that, per trip, only 4.44 € are saved. This becomes even clearer if, for a moment, we compare two cars with the same tank volume, 45 l, and the same consumption rate, 0.07 l/km, both in Trier with 1.19 l in the tank, the first one driving to Luxembourg to fill up and the second one driving around the corner to put only 42.62 l into the tank. In the end, both tanks contain 43.81 l, and the first driver has paid 53.10 € and the second 57.54 €.

We now check the recommendation in the solution of the filling up problem on a more fundamental basis. As mentioned, we have interpreted: "Is it worthwhile?" as "Do I save money?" However, is it appropriate to consider only the money issue? It takes time to drive to Luxembourg and back, perhaps a lot of time if there is a traffic jam on the road. Who is Mrs. Stein, and can she afford spending that time when she must fill up her car? If she can, how much could she have earned per hour for the time she spends? Furthermore, there is always a risk, albeit probably a small one, of having an accident on the trip, and there is additional air pollution caused by this trip which Mrs. Stein might wish to avoid. In addition, each car loses value with growing mileage, so we should include several more parameters in our calculations, not only the costs of filling up and of driving. Dependent on which aspects we choose to consider, the decision whether to drive to Luxembourg or not might turn out differently.

Now, we take one more parameter into account: the time that Mrs. Stein spends on her trip to the station in Luxembourg, thus aiming at a second solution. If the conditions (traffic density, weather) are normal on the trip to Luxembourg and Mrs. Stein lives not too far from the road to Luxembourg, we can assume an average speed of 70 km/h. This means that the trip altogether takes 34 km/(70 km/h) ≈ 0.5 h, that is, about 30 minutes. If Mrs. Stein's time is worth the German minimum wage of roughly 9 € per hour, this trip costs 4.50 € in terms of lost income, which is almost exactly the same amount of money she saves by undertaking the trip. If her wages are higher, then she actually invests more into this trip than the costs she saves. However, Mrs. Stein might like driving and might find the trip to Luxembourg a recreation from work, or she might have friends in Luxembourg that she can visit on the trip. Therefore, we cannot find an appropriate solution without knowing more about Mrs. Stein – a typical situation when real life meets mathematics.

When it comes to validating the model answers, it is interesting to observe that this cannot be done solely on empirical grounds. While it is possible to check the distance between Trier and the station in Luxemburg, the petrol prices in both places, as well as the capacity of the tank in Mrs Stein's car, the modelling conclusion concerning the cost saving obtained depends on some calculations involving the consumption rate. These require data on the distances actually travelled and the amounts of fuel actually consumed during the different sections of the trip. So, only *post hoc* checking of the model answers is possible. As to an evaluation of the model, some aspects have already been touched upon above. For instance, several significant factors for assessing the worthwhileness of the trip have deliberately been left out of consideration in the model described. These factors include time spent on the journey, potential loss of income resulting from spending time on driving rather than on earning an extra income, environmental considerations, etc. If factors other than cost savings are of importance, this model is not fully inadequate. Even with cost as the only focus, however, the model doesn't account for fuel consumption while waiting in a queue or at traffic lights. Moreover, the crucial parameters in the model are specified without paying attention to uncertainties. How would uncertainties in the specification of distance, tank volume, and fuel consumption influence the model outcome? Some of these issues are addressed in the generalised situations considered below.

A generalised solution

We now generalise the problem and its first solution by not only considering Mrs. Stein's car but any car in Trier with tank volume V and consumption rate C. Again, we take only the costs for filling up and for driving into account; no other variables such as time are considered, and we still assume that the station in Trier is just round the corner. Then the price for filling up in Trier is V · 1.35 €/l and for filling up in Luxembourg (if the whole trip is made with "Luxembourg petrol"):

V · 1.18 €/l + 34 km · C · 1.18 €/l = (V + 34 km · C) · 1.18 €/l.

It is cheaper in Luxembourg if and only if:

(V + 34 km · C) · 1.18 €/l < V · 1.35 €/l,

or equivalently (rearranging this inequality and rounding off) 34 km · C < 0.144 · V or (rounding off again) C km < 0.0042 · V or in summary C/V km < 0.0042.

For Mrs. Stein's car, we have C = 0.07 l/km and V = 45 l, hence C/V km ≈ 0.0016, so we see again that it is worthwhile driving to Luxembourg in terms of money. How high can the consumption rate be so that it is still worth going if the tank volume is 45 l? The condition is:

(C/45 l) km < 0.0042, that is (rounded off) C < 0.19 l/km.

If the consumption rate is less than 0.19 l/km (which for normal cars is certainly fulfilled), then the trip is worthwhile. On the other hand: How small can the tank volume be so that it is still worth driving to Luxembourg, provided that the consumption rate is 0.07 l/km? The condition is:

(0.07 l/km)/V km < 0.0042, that is (rounded off) V > 17 l.

Normal cars certainly have a tank volume bigger than 17 l. We can, of course, vary both quantities, C and V, simultaneously. If, only for a moment, we omit all units and measure C in l/km and V in l, then the condition for equal costs in Trier and Luxembourg is simply C ≈ 0.0042 · V, a proportional relation.

Another generalised solution

We have just generalized Mrs. Stein's car to any car while keeping all the other parameters unchanged. Another generalisation which is very natural is to vary the unit prices per litre for petrol. Let P_1 be the price in Trier and P_2 be the price in Luxembourg. To have only two variables to consider, we again take Mrs. Stein's car with a tank volume of 45 l and a consumption of 0.07 l/km. The cost (in Euros) of filling up in Trier is then 45 l · P_1 and the cost in Luxembourg:

(45 l + 34 km · 0.07 l/km) · P_2 = 47.38 l · P_2.

It is cheaper to go to Luxembourg if and only if:

47.38 l · P_2 < 45 l · P_1, or (rounded off) P_2 < 0.95 · P_1 or P_2/P_1 < 0.95.

In our example, we have P_2/P_1 ≈ 0.87. If the price in Trier is P_1 = 1.35 €/l, then the price in Luxembourg must be less than (rounded off) 1.28 €/l for it to be worth driving there. If the price in Luxembourg is P_2 = 1.18 €/l, then the price in Trier must be more than (rounded off) 1.24 €/l. This value could have been calculated in the beginning: Driving to Luxembourg costs 2.80 € and that must compensate for the higher petrol price in Trier, so the prices for filling up are equal exactly if the difference P of the petrol prices satisfies:

45 l · P = 2.80 €, thus P ≈ 0.06 €/l,

42 Modelling examples

which means a price of 1.24 €/l in Trier. If we vary both prices simultaneously, then the condition for equal prices in Trier and Luxembourg is $P_2 \approx 0.95 \cdot P_1$, again a proportional relation.

Yet another generalised solution

Finally, we can also vary the distances to the two stations (the distance to the nearby station was 0 km in all cases so far). If we assume again that the petrol in the tank is the same both ways, then only the distance between the stations is the relevant variable. Let D be that distance. To be specific, we include no more variables and take the data of Mrs. Stein's car as well as the same petrol prices as in the beginning. Then the difference of the prices for filling up is:

45 l · 1.35 €/l – (45 l · 1.18 €/l + 2D · 0.07 l/km · 1.18 €/l) = 7.65 € – D · 0.1652 €/km.

Leaving other factors out of consideration, it is worth driving to the distant station if and only if this difference in price is bigger than 0 €, which is equivalent to (rounded off):

D < 46 km.

A general mathematical model

Finally, we can set up an inequality where all relevant quantities are represented as variables. This can be interpreted as a general mathematical model of the given problem situation, again supposing we interpret "worthwhile" in terms of money only. If and only if the following inequality is true for given values of the variables, it is worth driving to the distant station:

$V \cdot P_1 - (V + 2D \cdot C) \cdot P_2 > 0.$

This inequality has a rather simple structure; it is linear in every individual variable and allows all the specialisations that we have carried out so far. The border case is when equality holds, that is, when the costs of filling up are equal in both places:

$V \cdot P_1 - (V + 2D \cdot C) \cdot P_2 = 0.$

We can now identify some variables that we deliberately consider as varying in this inequality respectively equation and others that we regard as parameters that are kept constant during such an analysis. For instance, if we regard P_1 and P_2 as variables and V, D and C as parameters, we get the functional dependence:

$P_2 = V \cdot P_1/(V + 2D \cdot C),$

which we looked at above for special values of V, D and C.

The equation also allows for functional considerations without thinking of the real context: What happens with . . . if . . .? An easy example: What happens with D_e (that is, the distance for which the costs for filling up here or there are equal) if P_1 increases (and all other variables are kept constant as parameters)? It is obvious (because of the " $-$ ") that D_e increases as well. The interpretation in the real world is: If the price in the nearby station gets higher, then the acceptable distance to the other station also gets higher – this is clear! What happens with D_e if P_2 increases (and the rest remain unchanged)? It is equally obvious that D_e decreases. What happens with D_e if P_2 increases by 5%? An increase by 5% means multiplication by 1.05, so the term $V+2D_eC$ is divided by 1.05 in order to maintain equality, which means $V+2D_eC$ decreases by circa 4.8%. The order of magnitude of V is somewhere in the range of 40 to 60 litres, C ranges between 5 and 12 litres per 100 km, and reasonable distances for D are below 30 km. That means the part $2D_eC$ in this term is at most 7 litres and thus significantly smaller compared to V. For the whole term $V+2D_eC$ to decrease by 4.8%, the smaller part $2D_eC$ and thus D_e has to decrease by much more than that. With P_1 = 1.35 €/l, P_2 = 1.23 €/l, V = 50 l, D = 25 km, C = 0.1 l/km we have approximate equality. If P_2 increases by 5% to 1.29 €/l, then the new distance which brings equality is $D_e \approx 12$ km, so the new acceptable distance is less than half of the former acceptable distance.

Concluding remarks

The purpose of the modelling carried out in this example was to pave the way for individuals, such as Mrs. Stein, to make decisions and take possible subsequent actions concerning the issue of whether or not it is worthwhile to drive to a distant petrol station to fill up the tank in one's car, instead of going to a nearby station that charges a higher price per litre. On the face of it, this may resemble a prescriptive modelling purpose because it may help shape the reality of an individual car user. However, the outcome of the modelling activity does not give rise to the kind of lasting design or organisation of physical, socio-cultural or scientific reality that is characteristic of prescriptive modelling. Rather, it offers an analysis of factors and conditions involved in covering and capturing the filling-up context, which is the purpose of descriptive modelling, in spite of the fact that the resulting model may actually be used as a tool for taking individual decisions or actions.

Example 3: Sight range

The problem: looking from a tower

Imagine we are standing on a tower, a high building or a mountain and want to know: How far can we see?

Let us take the Eiffel Tower in Paris as a concrete example (Figure 3.3).

44 Modelling examples

FIGURE 3.3 The Eiffel Tower in Paris

The upper part of the third platform, 279 m above ground, is accessible to visitors. Imagine we are standing on that platform. The surroundings of the site are more or less flat, meaning that there are no mountains in the way to block our vision.

Let us assume clear visibility and that we can use a telescope if we like so that there is no visual restriction on our view. If we look into the sky, we can certainly look arbitrarily far. By the question "How far can we see?" we actually mean a slightly different question, namely "How far away is the horizon from where we stand?" where the "horizon" consists of all points on the surface of the earth that are at a maximum distance from us among all visible points. The horizon is (approximately) a circle (the word "horizon" is of Greek origin, and its original meaning was "limiting circle"). If we make a plane cut through the earth, through its centre, and look at the situation from far away in outer space, we see the earth idealised as a circle as well as two points of the horizon determined by our position. Let us take one of these points, represent the sight beam from our eyes to this point by a line segment and connect both the Eiffel Tower and the horizon point with the midpoint of the earth (see Figure 3.4).

Now we can mathematise the situation by means of a circle and a triangle, as in Figure 3.5.

By h we denote the height of the tower, including the height of an observer (so we have $h \approx 281$ m), while s is the length of the sight beam (which is what we are seeking) and R is the radius of the earth (on average $R \approx 6,371$ km). The triangle with side lengths R, s and $R+h$ is right-angled since the sight beam lies on a tangent

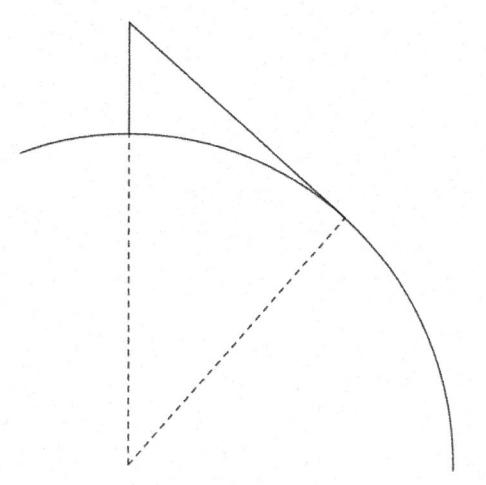

FIGURE 3.4 Cross section through the earth with Eiffel Tower and horizon point

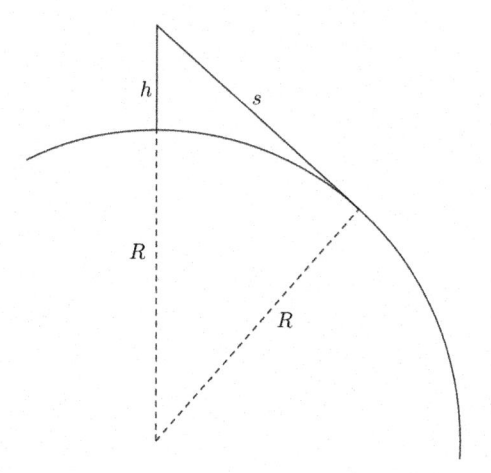

FIGURE 3.5 Mathematical model of the sight situation

line and tangents to a circle are perpendicular to its radius. We can use the Pythagorean theorem to obtain the following equation:

$R^2 + s^2 = (R + h)^2$,
whence $s^2 = 2Rh + h^2$ or
$s = \sqrt{2Rh + h^2}$.

By inserting quantities and rounding off to km, we get $s \approx 60$ km. The interpretation of this result is that the farthest we can see from the third platform of the Eiffel Tower is approximately 60 km.

Validation of the answer

How can we validate our answer? In principle, we could identify a certain object on the horizon and measure the distance from the tower to this object by means of appropriate instruments or by using GPS. We would find a non-negligible difference to our calculation – why? A less important reason is that the earth is not a complete sphere, so our model with a circle and the Pythagorean theorem is only approximately valid. The main reason, however, is the so-called refraction of the atmosphere. This means that light beams are not quite straight but slightly curved towards the surface of the earth, which then appears to be bigger than it is. In other words, one can see a little bit beyond the geometric horizon. The actual size of the refraction depends on temperature, humidity and air pressure. The so-called "apparent earth radius" due to refraction is on average $R_1 \approx 7,680$ km, so the real sight distance from the Eiffel Tower is $s_1 = \sqrt{2R_1 h + h^2} \approx 66$ km; this is about 10% larger than what was calculated from our first model. This is a rule of thumb that holds for all sight distances since R_1 is roughly 20% larger than R and because the square root of the earth radius is needed to determine the sight distance, s_1 is about 10% larger than s (as $\sqrt{1.2} \approx 1.1$).

In the following, we will, nevertheless, continue to use the first model because it does not require physical knowledge beyond usual, everyday knowledge.

Another critical reflection on our solution concerns the question of what we mean by "sight distance". Do we really mean the distance between the point where we stand on the tower and the horizon, or do we mean the distance on the surface of the earth between the foot of the tower and the horizon? Normally, distances between objects on the earth are measured on the surface, so, technically speaking, a distance is the length of the arc on the corresponding circle. Let us calculate this arc length in the present example. For that, we need the angle α at M in our right-angled triangle (see Figure 3.6).

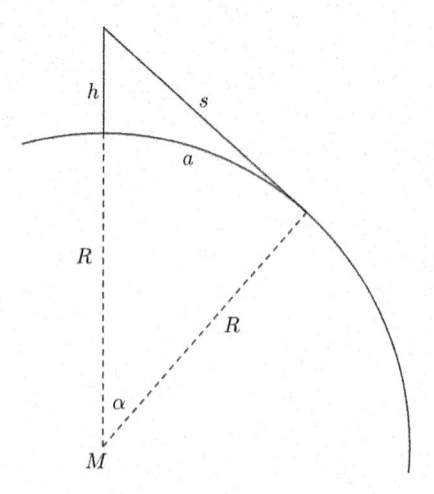

FIGURE 3.6 Mathematical model with angle

By means of trigonometry, we get $cos(\alpha) = R/(R+h) = 6{,}371/6{,}371.281 \approx 0.99995558$ and hence $\alpha \approx 0.54°$ – a very small angle! This is clear because h is very small compared to R, as is s, so the triangle in our model is very "flat". Now we can calculate the unknown arc length a via $a/C = \alpha/360°$ where $C = 2\pi R$ is the circumference of the earth circle. The result is, rounded off to integer kilometres, $a \approx 60$ km, the same result as in the first model. Therefore, the difference between the arc length and the distance in the air does not matter. This was to be expected since the height of the tower is very small compared to this distance, so the triangle formed by the tower, the sight line segment in the air and the sight arc on the earth are also very "flat".

A generalisation

We have solved the special problem of finding how far we can see from the third platform of the Eiffel Tower. However, the formula we have developed holds for any height h. So, for instance, if we look from the roof of a building which is 120 m high, the formula gives a sight distance of circa 39 km. We see in our calculation that the term h^2 only contributes marginally to the result (in fact, the two results with and without h^2 for $h = 120$ m differ by 0.00018 km; this is 18 cm – not a reasonable accuracy to take into account, given the accuracy of the rounded quantities R and h). As h is very small compared to R, h^2 is very small compared to hR and thus can be omitted. We end up with the formula $s \approx \sqrt{2Rh}$ where R is a constant. This means that s is (approximately) proportional to the square root of h, a remarkable result. For instance, if we want to see twice as far, we have to be four times as high up. Differently put, a tower twice as high as another tower allows for an approximately $\sqrt{2}$ times as big sight distance (i.e., 41% more).

If we measure h in m, say $h=h_0$ m where h_0 is a positive real number indicating the numerical value of the quantity h, we can calculate:

$$\sqrt{2Rh} \approx \sqrt{12{,}742\text{km} \cdot h_0\,\text{m}} = \sqrt{12.742 \cdot h_0\,\text{km}^2} \approx 3.57 \cdot \sqrt{h_0}\,\text{km}$$

or roughly $s \approx 3.5 \cdot \sqrt{h_0}$ km. In words: If you want to know your sight distance in km, take the square root of your distance from the ground in m and multiply this number by 3.5 (sometimes known as the "mountain climber rule"). With this formula, we can calculate approximate sight distances within seconds in our head. For instance, in the case of the Eiffel Tower, with $h = 281$ m, we have $\sqrt{h_0} = \sqrt{281} \approx 17$ and therefore $s \approx 3.5 \cdot 17$ km ≈ 60 km.

With this formula, we can also answer an inverse question quite easily. Supposed we are in the Forêt de Fontainebleau, 65 km away from the Eiffel Tower. Can we still see part of the tower? Dividing 65 by 3.57 and squaring the result, we get $h_0 \approx 330$, which means we can see, from this point, something which is approximately 330 m high. The highest point of the Eiffel Tower is 324 m; according to this calculation, we cannot see this point. However, due to refraction (see above), we might nevertheless see the top of the antenna on the Eiffel Tower.

The lighthouse problem

We can now treat a related but different problem in the context of sight beams and sight distances. In former times, lighthouses with their beacons showed ships that they were approaching the coast. As an example, we take the famous lighthouse Faro do Cabo de São Vicente on the Southern coast of Portugal. It is 28 m high and located on a rock so that its beacon is 86 m above sea level (Figure 3.7).

The question is: How far away from this lighthouse is a ship when people onboard can see the beacon of the lighthouse for the first time?

We presuppose again that the earth is a sphere (so, for instance, no water waves are influencing the sight) and that the weather conditions are good. If the ship is only considered as a point on the water, then this question is simply a special instance of the sight problem that we just solved (for it is, of course, the same distance looking from the lighthouse to the ship as vice versa). So the point-like ship is $3.57 \cdot \sqrt{86}$ km ≈ 33 km away. However, a ship is not a point but an object with a height. We imagine a ship where a sailor looking for the lighthouse is 15 m above sea level. Then the lighthouse can be seen for the first time before the ship reaches the horizon point from where the lighthouse would be seen for the first time from a point-like ship. Instead, it can be seen from the moment when the observer can see that horizon point. An appropriate mathematical model comprises two adjacent right-angled triangles, with sides on a common tangent line, as in Figure 3.8.

We denote by H the height of the lighthouse, by h the height of the ship, by S the sight distance from the lighthouse to the horizon point in direction of the ship, and

FIGURE 3.7 Lighthouse Faro do Cabo de São Vicente in Portugal

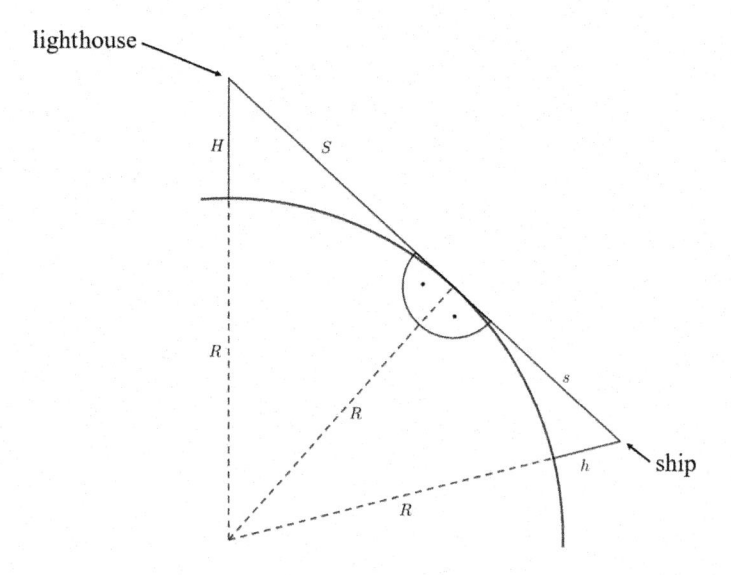

FIGURE 3.8 Mathematical model of the lighthouse situation

by s the sight from the ship to the same horizon point. We are looking for the distance between ship and lighthouse, which is $S + s$. Then, similar considerations and calculations as in the Eiffel Tower problem (Pythagorean theorem applied twice) result in $S \approx \sqrt{2RH}$ and $s \approx \sqrt{2Rh}$, so:

$$S + s \approx \sqrt{2RH} + \sqrt{2Rh} = \sqrt{2R} \cdot (\sqrt{H} + \sqrt{h}).$$

We had already calculated $S \approx 33$ km for $H = 86$ m, and we get $s \approx 14$ km for $h = 15$ m, so altogether we find circa 47 km. Thus, the answer to our question is: A ship with a height of 15 m is approximately 47 km away from the lighthouse at Cabo de São Vicente when the lighthouse can be seen for the first time.

Concluding remarks

The sight range problem exemplifies that mathematical modelling often requires real-world knowledge, in this case that the earth is approximately a sphere. In classroom experiments at school and university, the first approach by many students to solve this problem is a sketch of a right-angled triangle with the tower, a segment of the flat earth and the sight beam from the top of the tower to the other end of this segment. It is a non-trivial step to realise that this is an inappropriate model because the earth is not flat and what limits the sight range is the curvature of the earth.

The example also shows that sometimes an established model can be used and extended to cover a more complex problem situation. If we started with the lighthouse problem, we would have to develop first the same model as for the Eiffel Tower problem (one triangle) and then extend it (two triangles) again by using real-world knowledge about the earth. Another more general feature of this example is the possibility of different mathematical treatment approaches once the mathematical model has been established. The sight distance can be calculated by using either the Pythagorean theorem or trigonometric relations. A purely graphic approach (which often is an alternative to calculations when geometric shapes like triangles are involved) would not work here because one angle in the model triangle is too small to be accurately drawn.

Example 4: Paper formats (DIN A)

Designing paper formats

Let us first imagine that we wish to *design* a system of paper sheet formats with the following three properties (cp. Figure 3.9):

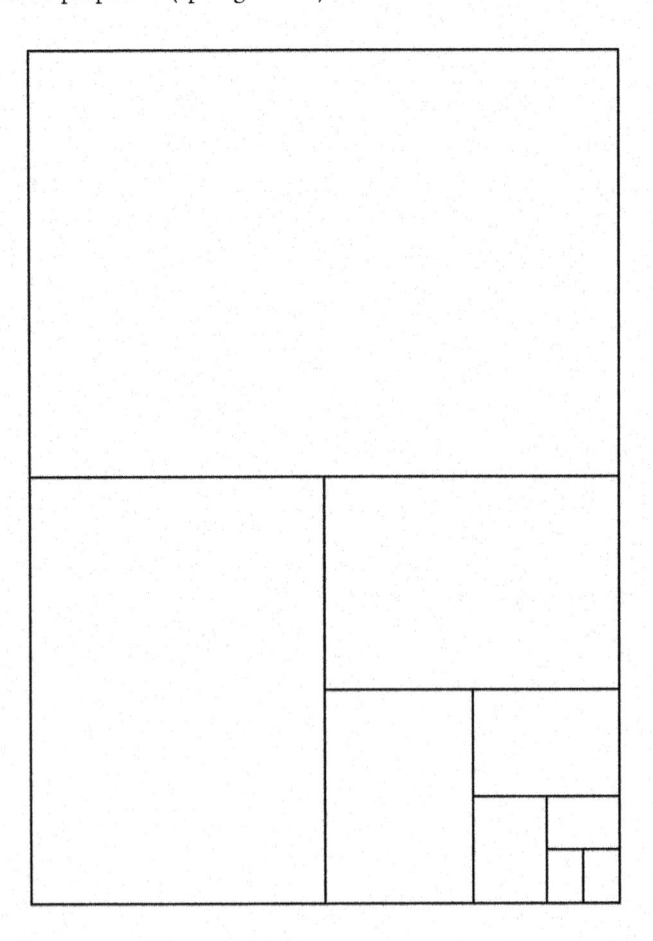

FIGURE 3.9 The system of A-paper formats

1   Each sheet of paper is rectangular.
2   The area of the largest sheet of paper in the system is 1m².
3   If any sheet of paper in the system is bisected across a mid-point transversal between the two longer sides, each half sheet also belongs to the system and is similar to the previous one, i.e., the proportions between corresponding sides are the same.

These properties reflect certain practical needs or advantages in dealing with such paper sheets, including aesthetic aspects (property 3 means that each sheet "looks the same" when seen from any distance). Major advantages of having this system are that any sheet in the system can be made from the same basic "mother sheet" by a number of simple cuts and that everybody dealing with paper sheets, be it everyday users or producers of envelopes, knows what sizes are available and can act accordingly.

But can these requirements actually be satisfied? If so, what are the dimensions of each sheet? Is there more than one solution to the design problem?

In addition to the requirements stated, dealing with this problem involves making a few basic assumptions about the paper world reality we have in mind: Sheets can be cut precisely by machinery and are thin enough to be folded at least once, as we wish.

We are now ready to mathematise the situation. Let us begin by observing that each sheet has the same shape, so we are dealing with a scaling problem. As each successive sheet is half that of the previous one, the length scales down by $2^{1/2}$.

Now for the details: We denote the n'th sheet in the system by A_n, $n \geq 0$. Corresponding to requirement (1), A_n is mathematised as a rectangle, defined in terms of its dimensions (l_n, s_n), where l_n indicates the length of the longer side and s_n the length of the shorter side.

The remaining requirements, (2) and (3), are mathematised as:

- (a) Largest sheet, A_0: $l_0 \cdot s_0 = 10{,}000$ cm²
- (b) Similarity: For every $n \geq 0$ we must have $l_{n+1}/s_{n+1} = l_n/s_n$,
- (c) Sheet bisection: $l_{n+1} = s_n$ and $s_{n+1} = l_n/2$, for every $n \geq 0$.

The mathematised questions then become:

- Does there exist a sequence of pairs (l_n, s_n), $n \geq 0$, consisting of positive elements, that satisfies (a), (b) and (c)?
- If so, what elements does/can this sequence of pairs have?

The mathematical domain for this mathematisation consists of lengths, that is, non-negative real numbers with length units (usually cm or mm) and sequences.

We are now ready to answer these questions by undertaking a mathematical treatment of the situation.

52 Modelling examples

First, we observe that since for all $n \geq 0$, $l_{n+1}/s_{n+1} = l_n/s_n$, and $l_{n+1} = s_n$ as well as $s_{n+1} = l_n/2$, we obtain by insertion of the latter relation into the former: $l_n/s_n = l_{n+1}/s_{n+1} = s_n/(l_n/2)$, which gives us an equation linking l_n and s_n, namely $l_n^2 = 2s_n^2$, i.e.,:

$$l_n = 2^{1/2} s_n,$$

valid for all $n \geq 0$.

This also holds for $n = 0$, yielding $l_0 = 2^{1/2} s_0$. Now, since we must also have $l_0 s_0 = 10^4$ cm^2, insertion of l_0 expressed in terms of s_0 yields $2^{1/2} s_0^2 = 10^4$ cm^2, i.e., $s_0 = 10^2/2^{1/4}$ cm. Hence, $l_0 = 2^{1/2} s_0 = 2^{1/2} 10^2/2^{1/4}$ cm $= 2^{1/4} 10^2$ cm. So for $n = 0$, we have obtained:

$$l_0 = 2^{1/4} 10^2 \text{ cm, and } s_0 = 10^2/2^{1/4} \text{ cm.}$$

Next, we have $l_1 = s_0 = 10^2/2^{1/4}$ cm and $s_1 = l_0/2 = 2^{1/4} 10^2/2 = 10^2/2^{3/4}$ cm. Continuing by recursion, we obtain for any $n \geq 0$:

$$l_n = 10^2/2^{(2n-1)/4} \text{ cm and } s_n = 10^2/2^{(2n+1)/4} \text{ cm,}$$

relations which can be formally proved by induction on n.

These results answer the initial questions. Yes, there does exist a – uniquely determined – (infinite!) sequence of paper sheet formats satisfying all the desired properties. The dimensions of sheet A_n are:

$$l_n = 10^2/2^{(2n-1)/4} \textbf{ cm and } s_n = 10^2/2^{(2n+1)/4} \textbf{ cm.}$$

Thus, for example, the dimensions of the prevalent A_4 sheet are:

$$l_4 = 100/2^{7/4} \text{ cm} \approx 29.7301778751 \text{ cm} \approx 29.7 \text{ cm and}$$
$$s_4 = 100/2^{9/4} \text{ cm} \approx 21.0224103813 \text{ cm} \approx 21.0 \text{ cm.}[1]$$

What we have done in this section is that we have helped create a piece of reality by designing a uniquely determined system of paper formats fulfilling certain initial requirements. More specifically, we have undertaken "prescriptive modelling".

How can we validate this model? The fundamental validation consists of observing that our initial three wishes allowed for a translation into mathematical requirements that could indeed be fulfilled – in fact in one and only one way. If more than one solution had existed, we could have decided to consider additional wishes or requirements, but as this is not the case, the only thing we can do is to re-visit these wishes and check, once again, whether they still correspond to the initial purposes of the whole endeavour or whether they ought to be subject to change or amendment. Beyond this, as the model is creating reality rather than describing it, no further confrontation of the model with reality is meaningful.

Analysing existing paper formats

However, the system of A-paper formats does, in fact, already exist as a physical and social reality. The papers can be purchased in millions of paper and stationery shops around the world. Imagine now that we wanted to uncover a possible underlying pattern in this system of paper formats which is taken to be unknown to us, thus asking the question: What are the patterns and principles in and behind these sheet formats?

Conducting an empirico-physical investigation of these paper formats would involve making physical measurements of the formats, leading to the discovery of the rectangular shape of all sheets (property 1) as well as to tables of lengths and widths of the sheets, which also can be retrieved from various official websites of A-paper formats. Here is an excerpt of such a table (all lengths and widths rounded off to whole millimetres):

A0: 1,189 mm × 841 mm
A1: 841 mm × 594 mm
A2: 594 mm × 420 mm
A3: 420 mm × 297 mm
A4: 297 mm × 210 mm
A5: 210 mm × 148 mm
A6: 148 mm × 105 mm
A7: 105 mm × 74 mm
A8: 74 mm × 52 mm
A9: 52 mm × 37 mm
A10: 37 mm × 26 mm

If we mathematise this situation by denoting, again, the dimensions of the longer and smaller sides by l_n and s_n, respectively ($n \geq 0$), we *observe* directly from the table that:

i $\quad l_{n+1} = s_n$ for $n = 0, \ldots, 9$.

It also looks as if:

$$s_{n+1} \approx l_n/2 \text{ for } n = 0, \ldots, 9,$$

which leads us to stating this as a *hypothesis*, that is:

ii $\quad s_{n+1} = l_n/2$ for $n = 0, \ldots, 9$.

Combining observation i and hypothesis ii, we obtain, as a derived hypothesis, that:

iii $\quad (l_{n+1}, s_{n+1}) = (s_n, l_n/2)$ for $n \geq 0$,

54 Modelling examples

which means that sheet $A(n+1)$ is assumed to be obtained by folding sheet An along the midpoint transversal of the longer sides, which corresponds exactly to the first part of property 3 above. From i and hypothesis ii, we immediately get:

$$l_{n+1}/s_{n+1} = s_n/(l_n/2) = 2s_n/l_n, n \geq 0$$

and therefore also:

$$l_n/s_n = 2s_{n-1}/l_{n-1}, n \geq 1,$$

which by insertion into the previous equation yields:

$$l_{n+1}/s_{n+1} = 2s_n/l_n, = 2 \, (l_{n-1}/2s_{n-1}) = l_{n-1}/s_{n-1}, \text{ for } n \geq 1.$$

From this relationship, we make the more general hypothesis that the ratios between the sides are the same for all sheets:

iv $\quad l_{n+1}/s_{n+1} = l_n/s_n$ for $n \geq 0$

which corresponds exactly to the second part of property 3 above. From these relationships, we can deduce in the same way as in the previous section that:

$$s_n/(l_n/2) = l_n/s_n,$$

whence

$$l_n = 2^{1/2} s_n.$$

Especially, $l_0 = 2^{1/2} s_0$, and $l_1 = 2^{1/2} s_1 = 2^{1/2} l_0/2 = 2^{-1/2} l_0$, and further on, yielding:

$$l_n = 2^{-n/2} \, l_0, \text{ for } n \geq 0.$$

Therefore, as $s_n = 2^{-1/2} l_n = 2^{-1/2} \, 2^{-n/2} l_0$, we also have:

$$s_n = 2^{-(n+1)/2} l_0, n \geq 0.$$

This still does not capture the specific dimensions of the paper sheets, only the relationships between them. However, we see that all the dimensions are specified in terms of the value of l_0, which we can read off from the official table, $l_0 = 1{,}189$ mm.

Unlike the previous situation, this time we can validate our model with respect to reality. We shall do this by confronting the model outcomes with real data (www.da.wikipedia.org – again all values rounded off to whole mm):

Reality

A0: 1,189 mm × 841 mm
A1: 841 mm × 594 mm
A2: 594 mm × 420 mm
A3: 420 mm × 297 mm
A4: 297 mm × 210 mm
A5: 210 mm × 148 mm
A6: 148 mm × 105 mm
A7: 105 mm × 74 mm
A8: 74 mm × 52 mm
A9: 52 mm × 37 mm
A10: 37 mm × 26 mm

Model outcomes

A0: 1,189 mm × 841 mm
A1: 841 mm × **595 mm**
A2: **595 mm** × 420 mm
A3: 420 mm × 297 mm
A4: 297 mm × 210 mm
A5: 210 mm × **149 mm**
A6: **149 mm** × 105 mm
A7: 105 mm × 74 mm
A8: 74 mm × **53 mm**
A9: **53 mm** × 37 mm
A10: 37 mm × 26 m

It is evident that the model outcomes are almost identical to the real values and in the 3 out of 12 cases in which there is a difference (marked in bold in the table above), this difference is very minor. So, it is fair to claim that the model outcomes have been validated with overwhelming success, so much that the model itself deserves a very positive evaluation. In other words, we can safely claim that our descriptive modelling activity has successfully uncovered the principles and mechanisms that underlie the data pattern observed.

We observe that $2^{1/2}s_0^2 = l_0 s_0 = 999,949$ mm$^2 \approx 10^6$ mm^2 when:

$$s_0 = 10^3 2^{-1/4} \text{ mm and } l_0 = 10^3 2^{-1/4} 2^{1/2} \text{ mm} = 10^3 2^{1/4} \text{ mm},$$

yielding:

$$l_n = 2^{-n/2} 10^3 2^{1/4} \text{ mm} = 10^3 2^{-(2n-1)/4} \text{ mm}$$

and:

$$s_n = 2^{-1/2}l_n = 2^{-1/2}10^3 2^{-(2n-1)/4} \text{ mm} = 10^3 2^{-(2n+1)/4} \text{ mm}.$$

Altogether, our mathematisation has re-generated properties 1, 2 and 3 above, as well as the mathematical consequences of these properties. So, the outcome of the descriptive modelling we have conducted is as follows:

$$l_n = 10^3 / 2^{(2n-1)/4} \text{ mm} = 10^2 / 2^{(2n-1)/4} \text{ cm}$$

and:

$$s_n = 10^3 / 2^{(2n+1)/4} \text{ mm} = 10^2 / 2^{(2n+1)/4} \text{ cm for } n \geq 0.$$

In other words, our descriptive mathematical modelling of the reality of A-paper formats has given rise to the very same results as obtained in the prescriptive modelling of the first section.

Concluding remarks

This example presents and analyses two sides of the same coin. First, we assumed that we wanted to create a piece of paper world reality that fulfilled some more or less natural wishes or requirements. We succeeded in realising, by mathematical means, that this is possible in one and only one way. Second, we took the dual view of considering and describing aspects of an already existing paper world to uncover the principles and mechanisms that lie behind the system of A-paper formats. We succeeded to do so to such an extent that if the design principles and mechanisms thus uncovered were put to use, this would result in the very same paper sheet formats (modulo insignificant deviations stemming from rounding) that came out of the prescriptive modelling activity presented in the first section. This might seem to be self-evident, but it is not since we arrived at the same result from two very different points of departure, a theoretical and an empirical one.

It should finally be mentioned that this example displays a duality between prescriptive and descriptive modelling, very similar to the duality in the loan amortisation example to follow.

Example 5: Loan amortisation

The problem situation: loans

In the financial world, there is a multitude of different types of loans to finance private investments, such as buying a home, a summer cottage, a car or a boat;

renovating your garden, roof, or heating system; or simply financing parts of your general consumption. One such type of loan is the so-called "constant repayment loan", also known in some countries as an "annuity loan". In such a loan – which one typically gets from a bank or another kind of financial institution – you enter an agreement with the provider of the loan, let's called it "the bank" for simplicity. You agree that you will have repaid a loan of a certain sum, called the "principal", plus "interest", after a certain number of terms, at each of which you pay a fixed amount of money, called the "service payment per term". The "interest rate" – per year or per term – is fixed throughout the "maturity" (duration) of the loan, i.e., for each term. The service per term, which then is constant in an annuity loan, is composed as a sum of the interest per term and the repayment per term of the principal. The debt remaining after the service of a term has been paid is called the (outgoing) "balance". This amount then becomes the ingoing balance for the next term. A table listing the service, the interest and the repayment per term for all the terms is called the "amortisation" profile of the annuity loan, the word amortisation suggesting that this is the way in which the debt is, step by step, being "put to death".

Here is an example of an amortisation profile of an annuity loan from a certain financial institution in Denmark, Bankinfo. The principal of the loan is DKK 100,000; the maturity of the loan is 10 years, with service every quarter (i.e., 4 · 10 = 40 terms altogether); and the annual interest rate is 3%. The profile does not take inflation into account, so all figures refer to the year at which the loan was established. If future inflation were to be considered, assumptions or estimates concerning future inflation rates would have been needed.

First comes a table (supplied by the bank on its webpage and translated by us) summarising the loan:

Total service per year	DKK 11,612.06
Service per term	DKK 2,903.02
Service per month	DKK 967.67
Interest rate per term in %	0.75
Total interest during the maturity	DKK 16,120.62

Then comes the amortisation profile (also supplied by the bank). The table has to be read as follows: At the beginning of each term, *primo* indicates the ingoing balance; at the end of the term, the constant service of DKK 2,903.02 is paid, leaving an outgoing balance (*ultimo*) of the ingoing balance minus DKK 2,903.02 to the next term. The service is composed as a sum of an interest part and a repayment part, both of which vary along with the terms. Observe that numbers are rounded off to integer DKK:

58 Modelling examples

Year	Term	Primo Debt	Interest	Repayment	Ultimo Debt	Year	Term	Primo debt	Interest	Repayment	Ultimo Debt
1	1	100,000	750	2,153	97,847	6	1	53,729	403	2,500	51,229
1	2	97,847	734	2,169	95,678	6	2	51,229	384	2,519	48,710
1	3	95,678	718	2,185	93,492	6	3	48,710	365	2,538	46,173
1	4	93,492	701	2,202	91,291	6	4	46,173	346	2,557	43,616
1, sum			2,903	8,709		6, sum			1,499	10,113	
2	1	91,291	685	2,218	89,072	7	1	43,616	327	2,576	41,040
2	2	89,072	668	2,235	86,837	7	2	41,040	308	2,595	38,445
2	3	86,837	651	2,252	84,586	7	3	38,445	288	2,615	35,830
2	4	84,586	634	2,269	82,317	7	4	35,830	269	2,634	33,196
2, sum			2,638	8,974		7, sum			1,192	10,420	
3	1	82,317	617	2,286	80,031	8	1	33,196	249	2,654	30,542
3	2	80,031	600	2,303	77,278	8	2	30,542	229	2,674	27,868
3	3	77,278	583	2,320	75,408	8	3	27,868	209	2,694	25,174
3	4	75,408	566	2,337	73,071	8	4	25,174	189	2,714	22,460
3, sum			2,366	9,246		8, sum			876	10,736	
4	1	73,071	548	2,355	70,716	9	1	22,460	168	2,735	19,725
4	2	70,716	530	2,373	68,343	9	2	19,725	148	2,755	16,970
4	3	68,343	513	2,390	65,953	9	3	16,970	127	2,776	14,194
4	4	65,953	495	2,408	63,545	9	4	14,194	106	2,797	11,398
4, sum			2,086	9,526		9, sum			550	11,062	
5	1	63,545	477	2,426	61,118	10	1	11,398	85	2,818	8,580
5	2	61,118	458	2,445	58,673	10	2	8,580	64	2,839	5,741
5	3	58,673	440	2,463	56,211	10	3	5,741	43	2,860	2,881
5	4	56,211	422	2,481	53,729	10	4	2,881	22	2,881	0
5, sum			1,797	9,815		10, sum			215	11,398	

Now the question arises: Where do all these figures come from, i.e., how did the bank calculate them? Well, some of the figures in the summary table are easily traceable. The quarter term interest rate has been obtained by simply taking one-fourth of the annual interest rate of 3%. It is also clear that if we know the monthly service of DKK 979.67, we can find the quarterly service $979.67 \cdot 3 \approx 2,903.02$; the annual service $979.67 \cdot 12 \approx 11,612.06$; and the total service during the maturity of the loan by multiplying this number by 10, yielding 116,120.60. Subtracting the principal 100,000, we get the total interest paid during the 10 years, 16,120.60, which is almost the same as the figure indicated in the table. However, where on earth do the figure 979.67 and all the other figures in the amortisation table come from? In other words, what are the underlying principles and patterns that are responsible for generating all these numbers? It is worth noting that in the first terms of the amortisation period, the constant service includes a rather large interest, whereas in the final terms only a very small part of the service is interest.

Modelling the situation

The following are defining facts of annuity loans that have already been mentioned:

- One and only one amount of money, named the principal, was borrowed when the loan was established. We denote the principal amount by P, which is a positive real number with some currency unit attached.
  - In our example above, the principal is $P = 100{,}000$ DKK.
- The loan is to be paid back over a given integer number of terms, named the maturity of the loan. We denote the number of terms by the positive integer N.
  - In our example, each term is a quarter of a year, and the maturity of the loan is 10 years, hence 40 terms, i.e., $N = 40$.
- The interest rate per term is fixed during the maturity of the loan. We denote this rate by the positive real number r.
  - In our example, the interest rate per term is 0.75%, i.e., $r = 0.75/100$.
- At the end of each term, the borrower pays a fixed amount, named the service, to the provider of the loan. We denote this amount by S, which again is a positive real number with the same currency.
  - In our example, the service per quarter term was given as $S = 2{,}903.02$ DKK.

As we have seen, this is not enough to explain the specific amortisation profile of the loan, nor the way in which the service S has been determined. To provide answers to our questions we make some further assumptions:

1. At the end of each term, the interest generated by the ingoing balance, i.e., the remaining debt at the beginning of the term, is paid from the service. What remains of the service is used as repayment to reduce the balance of the term to generate a lower outgoing balance, which then is the new ingoing balance for the next term. This scheme is continued for all the terms, except the last one, from which there is no outgoing balance.
2. After the last service, at the end of the last term, the remaining debt is 0 *DKK*, and the loan has thus been amortised, i.e., paid fully back.
3. No inflation (or deflation, for that matter) is being taken into account.

We are now ready to mathematise the situation by translating the defining loan characteristics and the assumptions into mathematical terms.

After the end of the first term, the principal P has generated an interest rP, to be paid from the service S. The remainder $S - rP$ is used as repayment on the principal. This means that the remaining debt, the outgoing balance, P_1, after term 1 is given by:

$$P_1 = P - (S - rP) = P(1 + r) - S.$$

60 Modelling examples

We can also interpret this relationship as showing – equivalently – that the ingoing debt is increased during the term by adding the interest and then reduced by subtracting the service payment at the end of the term.

For term 2, the same pattern is repeated but now with P_1 as the new ingoing balance. So, at the end of term 2, the balance has been reduced to yield:

$$P_2 = P_1(1 + r) - S = (P(1 + r) - S)(1 + r) - S = P(1 + r)^2 - S(1 + r) - S$$

as the outgoing balance after term 2 and hence as the ingoing balance for term 3. Continuing in this manner throughout the terms, we obtain the following expression of the remaining debt, the outgoing balance, after term n, $n > 1$:

$$P_n = P_{n-1}(1 + r) - S = P(1 + r)^n - S(1 + r)^{n-1} - \ldots - S(1 + r) - S = P(1 + r)^n - S((1 + r)^{n-1} + \ldots + (1 + r) + 1).$$

Here we observe that the sum $(1 + r)^{n-1} + \ldots + (1 + r) + 1$ is the sum of a finite geometric series, which yields:

$$(1 + r)^{n-1} + \ldots + (1 + r) + 1 = [(1 + r)^n - 1]/[(1 + r) - 1] = [(1 + r)^n - 1]/r.$$

Inserting this, we obtain:

(i) $\quad P_n = P(1 + r)^n - S[(1 + r)^n - 1]/r$

as the general expression of the outgoing balance after term n.

This expression is derived on the assumption that S is known. However, it is not – yet! But we can in fact determine it by invoking assumption 2, that the loan is paid back after the end of term N. For, this assumption implies the equation (in the unknown S):

$$0 = P_N = P(1 + r)^N - S[(1 + r)^N - 1]/r,$$

such that:

(ii) $\quad S = [P(1 + r)^N \cdot r]/[(1 + r)^N - 1].$

Traditionally, in order to avoid powers both in the numerator and the denominator, this is reduced by multiplying both the numerator and the denominator by $(1 + r)^{-N}$, yielding:

(iia) $S = P\,r/[1 - (1 + r)^{-N}].$

Let us check whether this corresponds to the value of S (= 2,903.02 DKK) in our example, recalling that P = 100,000 DKK, r = 0.75/100, and N = 40. Utilising that $(1 + 0.75/100)^{-40} = 1.0075^{-40} \approx 0.741647961$, we get:

$S \approx 100{,}000 \text{ DKK} \cdot 0.0075/(1 - 0.741647961) =$
750 DKK$/0.258352039 \approx 2{,}903.37$ DKK,

which is very close to, but not completely identical with, the figure in the example. The difference is due to different rounding procedures in our computations and in those of the bank. Finding out who is right is not our business at this point. The most important thing is that we have been able to uncover the principles and the patterns of the amortisation profile of annuity loans.

We can further determine the other variables in the profile on the basis of (i) and (ii). The remaining figures that interest us follow immediately from the value of S.

Interest rates by term and by year

One more remark is warranted here. In the table given by Bankinfo, it is stated that an interest rate of 3% per year corresponds to an interest rate of 0.75% per quarter term. In fact, this is not correct if we refer to the usual way in which money grows in value on a bank account because the quarter term interest rate should give rise to compound interest during a year. If an amount A accumulates interest at a rate of r per quarter term, then A has increased to $A(1 + r)^4$ after four terms. If the annual interest is s, then we must have $A(1 + r)^4 = A(1 + s)$, which yields $s + 1 = (1 + r)^4$ or, equivalently, $r = (s + 1)^{1/4} - 1$. In the case of our example, where $s = 0.03$, we obtain $r = 1.03^{1/4} - 1 \approx 0.007417072$ and not 0.0075. This would give rise to a different value of S, namely 750 DKK$/(1 - 0.007417072^{40}) \approx 750$ DKK$/(1 - 0.744093908) =$ 750 DKK$/0.255906092 \approx 2{,}930{,}76$ DKK. This suggests that in the table provided by Bankinfo, the fixed point is the quarter term interest rate of 0.75% rather than an annual interest rate of 3%. Conversely, a quarterly interest rate of 0.75% corresponds to an annual interest rate of 3.0339% since $1.0075^4 - 1 \approx 0.30339191$.

However, the scheme adopted by many banks is that money grows with compound interest only if the time interval is more than one year. Within a year, banks oftentimes regulate the growth of money in a linear rather than in an exponential fashion, as did Bankinfo in our example, so that a quarterly interest rate of 0.75% without compounding does in fact correspond to an annual interest rate of 3%.

Validation and evaluation

Validating the model answers obtained consists in assessing how well these answers conform to the empirical reality. The empirical reality here is constituted by the amortisation profile presented by Bankinfo. Of course, other empirical realities stemming from other financial institutions might have been considered as well. In this case, the model answers are in very close agreement with the empirical amortisation profile, which means that we can consider the model answers validated.

Answer validation is one of the means by which the model built can be validated. As the purpose of the model is a descriptive one, namely to uncover the principles and assumptions underlying the empirical amortisation profile and to

62 Modelling examples

reconstruct the pattern displayed therein, it is fair to say that the model constructed does this job to a very satisfactory extent. In other words, the evaluation of the model yields a positive result. Of course, if we wanted the model to also take inflation into account – which was not included in the amortisation profile presented – the situation would have been different, as a more elaborate model would have been called for.

Concluding remarks

What we have presented here is a piece of descriptive modelling. Taking the point of view of a concerned consumer, or a financial analyst, we wanted to capture and understand a section of financial reality as represented by the amortisation table found on the internet. Basing our model on the assumptions listed in 1, 2, and 3 above, we were successful in this endeavour.

However, imagine that we had taken the point of view of a financial institution that wanted to construct a particular type of loan with the properties listed in 1, 2 and 3, which would then be design requirements, the very same amortisation profile would be obtained as a result of prescriptive modelling.

This example illustrates how different modelling purposes may lead to the same model, as was also discussed in 2.5.

Example 6: Traffic flow

The problem and various mathematical models

Imagine dense traffic on a long single-lane road. The problem we want to consider is this: At which speed should cars go to maximise the flow rate, in other words, to maximise the number of cars passing by per unit time? An obvious first answer seems to be "as fast as possible", but the faster the cars go, the larger the distance between two cars has to be for reasons of safety, and it is not obvious what an optimal balance between speed and safe distance would be.

To be accessible, the situation has to be simplified and structured. A natural simplifying assumption is:

1   All cars drive at the same speed (as they must when traffic is heavy),

supposing that we have a steady state traffic flow. Two other related and also rather obvious assumptions are:

2   All cars have the same length.
3   The distance between two cars is the same everywhere.

We write v for the speed of the cars, l for the car length and d for the distance between two cars (see Figure 3.10).

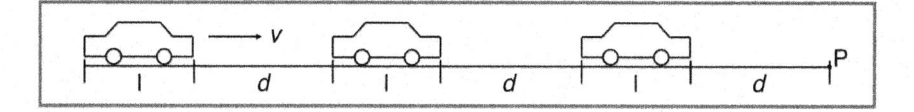

FIGURE 3.10 Situation picture for traffic flow problem with variables

It is clear that d somehow depends on v and that it does so in in a strictly isotonic way, that is, the larger v is, the larger is d. If we think of the real situation where two consecutive cars will have some distance between them and the driver in the car behind will have to adjust her/his speed to maintain a safe distance, it would also make sense to consider the inverse situation, where v depends on d. In the following, however, we shall always consider d as dependent on v.

Now, what exactly could/should "traffic flow rate" (sometimes also called "road capacity") mean? To answer this question, we imagine a fixed point on the road (point P in Figure 3.10) where someone sits and counts the passing cars, and we define the "traffic flow rate" F as the number of cars per unit time passing by this point. Let us illustrate the situation by a concrete example: If the cars all drive at 40 km/h and we count for one hour, then the following is clear: The last car which passes the counting point within this hour is at the beginning of the counting 40 km away, and exactly that number of cars now driving on this 40 km section of the road are counted. How many cars are there on this 40 km section? The answer is obvious by using division: Each car covers $l + d$, so there are 40km/($l + d$) cars on these 40 km. The flow rate is therefore this number divided by 1 hour, hence:

$$F = (40\,\text{km}/(l + d))/1\text{h} = (40\,\text{km/h})/(l + d) = v/(l + d).$$

The same considerations can be made with any speed. The result is: The appropriate mathematical model for the traffic flow rate is given by the formula:

$$F = v/(l + d),$$

where F, the number of cars per time, can be expressed in "number of cars/hour".

Now we must determine the distance between two cars, recalling that d is supposed to depend on v. There are various possibilities for "distance rules" that specify how d depends on v. The possibilities include the following rules, which in many parts of the world are actually used or even mandatory:

1 "Half speed rule": $d = (v_0/2)$ m (meters) where v_0 means the number of km/h of the speed (so, e.g., $v_0 = 40$ if $v = 40$ km/h) – a popular rule in many countries, although a rule which does not correspond to the physics of braking.
2 "1.5 seconds rule": $d = v \cdot 1.5$ sec (since speed is distance divided by time) – also a popular rule; for higher speeds often changed into the "2 seconds rule", equally ignoring the physical laws.

64 Modelling examples

3. "Driving school rule": $d = (3 \cdot v_0/10 + [v_0/10]^2)$ m (in Germany, one learns in the driving school "Divide the absolute measure of the speed in km/h by 10, multiply this number by 3 and also square it, and then add those two numbers; this gives you the distance you should keep in meters"); this distance model takes into account that the braking distance is in fact quadratically dependent on the speed and not linearly as the first two models presuppose.
4. "Stopping distance rule": $d = v \cdot t_R + v^2/(2 \cdot a)$ where t_R is the driver's reaction time (e.g., 1 second) and a is the braking deceleration (usual values in the range from 6 m/s² to 9 m/s²); this model, a generalization of the driving school model, assumes that it ought to be possible to brake if the car in front stops immediately.
5. "Front car rule": $d = v \cdot t_R + 1/2[1/a - 1/b]v^2$ where b is the braking deceleration of the car in front; this model takes into account that the car in front also needs some space to stop, so the model seems more appropriate for the actual situation.

In each case, the flow rate F is a function of the speed v, where the concrete function depends on the distance rule used: $d = d_i(v)$ ($i = 1, \ldots, 5$). So:

$$F = v/(l + d_i(v)) = f_i(v) \ (i = 1, \ldots, 5).$$

This implies that we have five different mathematical models of our traffic flow situation. The mathematical problem is then to maximise f_i. The first two distance functions are linear (considering only the reaction time); the next two are quadratic. The "front car" function is quadratic unless $a = b$; in this case, it is also linear (if both cars have the same braking deceleration then only the reaction time matters).

Let us ignore, for a moment, all the units and write down the five functions f_i with v as a variable (where v is measured in m/s), specifying l to 4 m, t_R to 1 sec, a to 6 m/s² and b to 9 m/s²:

$f_1(v) = v/(4 + v/2)$
$f_2(v) = v/(4 + 1.5v)$
$f_3(v) = v/(4 + 0.3v + v^2/10)$
$f_4(v) = v/(4 + v + v^2/12)$
$f_5(v) = v/(4 + v + v^2/36)$

Figure 3.11 shows the graphs of these five functions.

Solutions to the problem

We will now get different solutions to the original problem, depending on the mathematical models chosen. We can find easily that both f_1 and f_2 are strictly increasing and converging towards an asymptotic (saturation) value, 2 for f_1 and

Modelling examples **65**

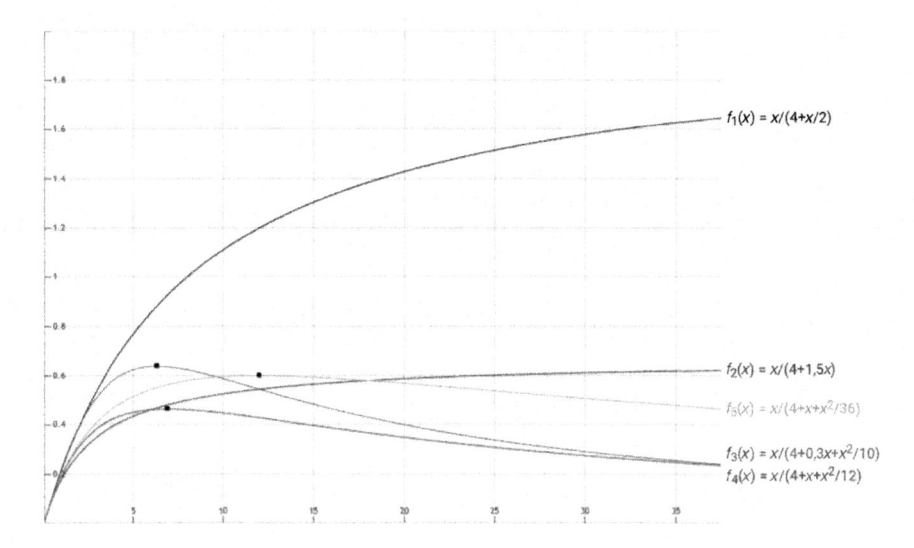

[Graph showing curves: $f_1(x) = x/(4+x/2)$, $f_2(x) = x/(4+1,5x)$, $f_5(x) = x/(4+x+x^2/36)$, $f_3(x) = x/(4+0,3x+x^2/10)$, $f_4(x) = x/(4+x+x^2/12)$]

FIGURE 3.11 Graphs of flow rate functions

2/3 for f_2. Both graphs are parts of hyperbolas, which can be seen by rewriting, for example, f_1 in the form:

$$f_1(v) = 2v/(8 + v) = 2(8 + v - 8)/(8 + v) = 2(1 - 8/(8 + v)) = 2 - 16/(8 + v)$$

so the graph of f_1 results from the graph of the basic hyperbola given by $f(v) = 1/v$ (only its right branch is relevant here) by a translation by 8 to the left, then a dilatation by 16 parallel to the second axis, then a reflection on the first axis and finally an upward translation by 2. When interpreted in terms of the real situation, this implies that in fact the flow rate increases with increasing speed in these two models, a result that passionate fast-drivers will appreciate. However, because of the asymptotic behaviour of f_1 and f_2, a relatively big increase of the speed will result in a relatively small increase of the flow rate only, so simply increasing the speed may not be worth the effort.

We further find that there exists a maximum value somewhere for each of the functions f_3, f_4 and f_5 since in all three cases it is clear, both formally and contextually, that $f_i(0) = 0$ and $f_i(v) \to 0$ if $v \to \infty$ and the functions are rational and hence continuous, that is, they make no leaps. The maximum value for these three functions can be determined, approximately, graphically or numerically, and precisely with methods from differential calculus. As an example, let us do the latter for f_4. The derivative of f_4 is:

$$f_4'(v) = [(4 + v + v^2/12) - v(1 + v/6)]/(4 + v + v^2/12)^2 = (4 - v^2/12)/(4 + v + v^2/12)^2.$$

So $f'_4(v) = 0$ exactly for $v^2 = 48$, and the only relevant solution is $v = \sqrt{48} = 4\sqrt{3}$. Therefore, f_4 has a unique maximum at $v = 4\sqrt{3}$. In a similar way, we find the maximum of f_3 at $v = 2\sqrt{10}$ and the maximum of f_5 at $v = 12$.

By the way, there is also a more elementary – i.e., "calculus free" – method for determining the maxima of these three functions exactly, but it requires some effort. All three functions are of the form:

$$f(v) = v/(1 + rv + sv^2), \text{ so } 1/f(v) = 1/v + r + sv, \text{ for } v \neq 0.$$

Instead of maximising f, we can minimise $1/f$ or (since r as a constant is irrelevant) minimise $g(v) = 1/v + sv$. The minimum of the sum of a proportional function and a reciprocal function is² situated where these two functions have the same value. So the minimum of g and hence the maximum value of f is at v with $1/v = sv$ that is $v = \sqrt{1/s}$, independent of r (which is interesting in itself). For the three functions this yields:

Minimum of f_3 at $v = \sqrt{40} = 2\sqrt{10}$
Minimum of f_4 at $v = \sqrt{48} = 4\sqrt{3}$ or more generally (see above) $v = \sqrt{2al}$
Minimum of f_5 at $v = \sqrt{144} = 12$ or more generally $v = \sqrt{2abl/(b-a)}$.

So far, we have ignored the units. If we insert the correct units and assume once again $l = 4$ m, $a = 6$ m/s² and $b = 9$ m/s² then the maximum of f_4 means, interpreted in the real-world situation, that the optimal speed for the cars according to the "stopping distance rule" is $\sqrt{48}$ m/s, that is approximately (since $\sqrt{48} \star 3.6 \approx 24.9$) 25 km/h – a remarkably low value! The maximum value for f_5 means that the optimal speed for the cars according to the "front car rule" is 12 m/s, that is approximately 43 km/h. This is also a value that passionate fast-drivers won't like but higher speeds will either reduce the flow rate or (if the distance rules are ignored) increase the risk of a rear-end collision – and in case of an accident, the flow rate would immediately be reduced to zero.

Concluding remarks

The final activity is to validate the results and, if need be, to refine the model and the results. First, we observe that the maxima of these three functions f_3, f_4, f_5 are "flat" as can be seen in Figure 3.11 (a property which maxima of differentiable functions generally have per se if they are continuously differentiable in a neighbourhood of the maximum point). This can be interpreted to saying that in reality it does not really matter for the flow rate if v is a bit smaller or larger than the precise maximum point.

There are other variables that we have not considered so far, e.g., the petrol consumption of the cars. The flat maximum just mentioned could again be interpreted as implying that it does not really matter for the flow rate if v is a bit smaller or larger than the precise maximum if this reduces the petrol consumption of the car. Another variable of significance to the problem is the reaction time at braking, which may also depend on the speed (higher speed may mean a higher stress level, which may influence the driver's attention in either direction) and, of course, on the attentiveness and the mental flexibility of the driver. Moreover, the whole problem depends on the technical equipment of contemporary cars. If all cars had optimal ABS brakes, then proportional distance rules would make more sense since in such circumstances the reaction time would be the decisive variable.

Example 7: Bertrand's paradox

Thought experiment

Imagine the following thought experiment, which was initially introduced by the French mathematician Joseph Bertrand (1889). An extremely thin rod of length l is being dropped at random onto a plane surface in such a way that it hits an extremely thin circular ring of radius r lying in the plane in two points (where r is smaller than $l/2$ so that the rod cannot lie entirely inside the circle), thus giving rise to a chord "cut off" from the rod (see Figure 3.12). What is the probability that the length of this chord is larger than the side of the inscribed equilateral triangle in the circle? Even though this may seem to be an artificial situation, it can actually correspond to a variety of physical experiments (see below) as well as computer simulations (Tessier, 1984).

Modelling the experiment

We can model this idealised experiment in three different ways. To do so, we prepare ourselves by considering a little elementary geometry (Figure 3.13).

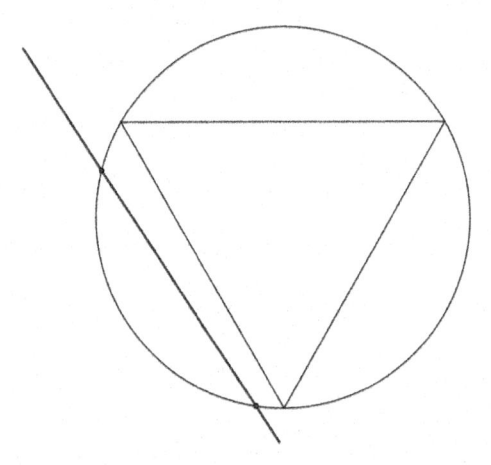

FIGURE 3.12 Rod, circle and inscribed equilateral triangle

68 Modelling examples

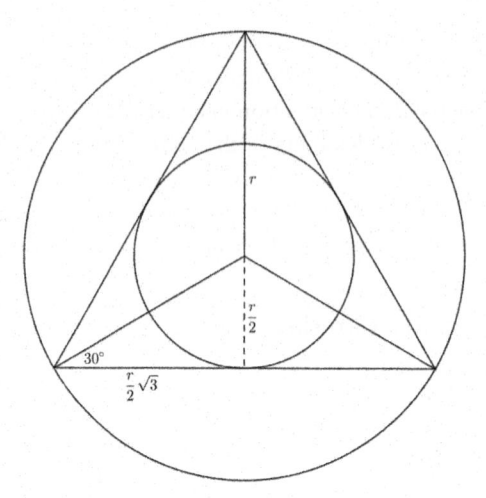

FIGURE 3.13 Measures in the inscribed triangle

The size of all the three angles of an equilateral triangle inscribed in a circle of radius r is 60^0. As the three radii from the centre of the circle going to the three vertices of the triangle bisect these angles, they form angles of 30^0 with the sides of the triangle. Hence, the size of the altitude of a smaller triangle formed by two vertex radii and the side joining the vertices is given by $r \sin 30^0 = r/2$. Similarly, half of the side length is given by $r \sin 60^0 = \sqrt{3}/2 \cdot r$, so the length of the side is $r\sqrt{3}$. Moreover, the equilateral triangle has an inscribed circle (with the same centre). As its radius is exactly the altitude of the small triangles determined by two vertices and the centre of the circle(s), the radius of the inscribed circle is $r/2$. One consequence of this is that the side of the inscribed equilateral triangle bisects the radius perpendicular to the side.

Model 1: A chord is given by two points on the circle, P and Q. Let us choose one of them, say P. We can consider the inscribed equilateral triangle with one of its vertices placed in P (see Figure 3.14). Then the length of the chord is larger than the length of the side in the triangle exactly if the other point Q is situated on the arc between the two other vertices of the triangle. Since that arc is 120^0, the probability that Q lies on it is $120^0/360^0 = 1/3$. So, in model 1, the probability that the length of the chord is larger than the length of a side of the inscribed equilateral triangle is $1/3$.

Model 2: For a randomly chosen chord, consider the inscribed equilateral triangle whose side is parallel to the chord (see Figure 3.15). The perpendicular bisector through the midpoint of the chord is also the perpendicular bisector of the triangle side at issue. It passes through the centre of the circle and hence gives rise to a radius in the circle. The length of the chord is larger than the length of the side exactly if the point of intersection between the chord and the radius under consideration is closer to the centre than is the point S of intersection between the side and the

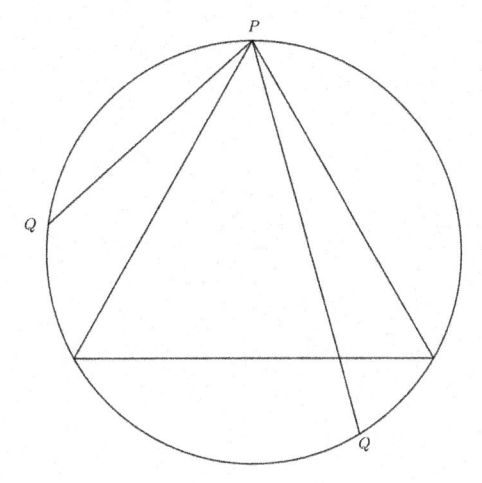

FIGURE 3.14 Chords in model 1

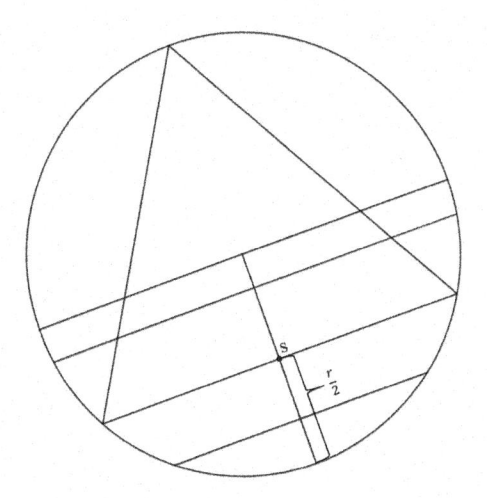

FIGURE 3.15 Chords in model 2

radius. Since the side of the triangle bisects the radius, the chord is longer than the side exactly if its intersection point with the radius lies on the interior half of that radius. The probability of a point on the radius lying on its interior half is 1/2. Hence, in model 2, the probability that the length of the chord is larger than the length of the side is 1/2.

Model 3: A chord in the circle is determined by its midpoint (except if the chord is a diameter). The length of the chord is larger than the length of a side of the inscribed equilateral exactly if this midpoint lies inside the disc corresponding to the inscribed circle of the equilateral triangle. Since this inscribed circle has a radius of 1/2 (see Figure 3.16), the probability of the midpoint lying in this circle

70 Modelling examples

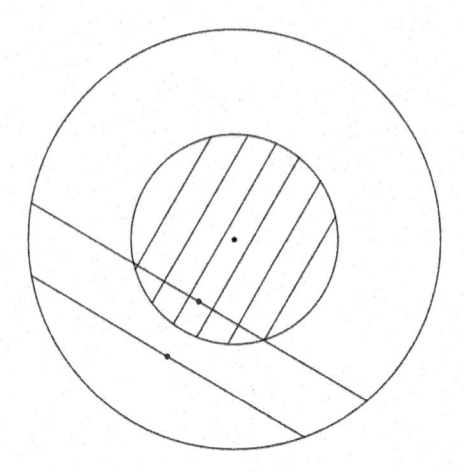

FIGURE 3.16 Chords in model 3

is 1/4 because the area of the smaller circle is 1/4 of the area of the larger circle. This is not affected by discarding the chords that are diameters since this amounts to removing the centre from the two areas, and the area of a set consisting of a single point is 0. So, in model 3, the probability that the length of the chord is larger than the length of the side is 1/4.

Since the three models are meant to capture the same random experiment, it is somewhat mind boggling that they yield three different results. Therefore, the situation described is known under the name of Bertrand's Paradox. Before looking into ways to settle this issue, we observe that the three models represent the randomness of the idealised experiment in three different ways.[3]

Analysing the models

In model 1, a random chord is specified by two points, P and Q, on the circle. As it only matters where Q is located once P is chosen, we can fix P in an arbitrary position on the circle and let Q vary randomly over the circle. Since Q can be located anywhere on the circle (except in P), we consider the positions of Q as uniformly distributed over the circumference of the circle. As the positions of Q that give rise to a chord longer than the side of the equilateral triangle are exactly those on the arc between to two vertices opposite of P, our result follows from the fact that the size of that arc is 1/3 of the size of the whole circle. In technical terms, the sample space involved consists of the points on the circle, represented as angles in the interval $[0, 360]$, and the probability distribution chosen is the uniform distribution on that interval, corresponding to a density function constantly equal to 1/360 on the interval.

In model 2, a random chord (except for diameters) is specified and represented by its midpoint. As the only thing that matters in determining whether the chord is longer than the side of the inscribed equilateral triangle is whether the midpoint of the chord is closer to the centre than half the radius, it is assumed in this model that all chords are placed in parallel so that their midpoints are all points on the same radius. We then assume those midpoints to be uniformly distributed over that radius, which can then be considered our sample space. This can be translated into a uniform probability distribution of distances to the centre over the interval [0, r] (corresponding to the density function constantly equal to 1/r on that interval).

In model 3, a random chord (except for diameters) is specified and represented by its midpoint. Once again, the length of the chord is larger than the side in the inscribed equilateral triangle exactly if its midpoint is closer to the centre of the circle than half the radius. Here, however, we take all possible positions of a chord into account, so the midpoint can be situated anywhere within the larger disc. So, we assume a uniform distribution of the midpoints over that disc, which is then considered our sample space. The chords which are longer than the triangle side are those with their midpoints positioned within the smaller disc of half the radius of the larger disc. The uniform probability distribution over the sample space is given by the density function constantly equal to $1/(pr^2)$ on the larger disc.

Why is it that we can establish three (and even more) reasonable models of the situation in the thought experiment and obtain three very different results? That is because the random thought experiment is not well defined (Kac, 1984). What exactly does it mean that a chord is dropped at random onto a circle in a plane? The three different models conceptualise such a dropping in three different ways, where the randomness is expressed in terms of three different (one-dimensional) uniform distributions over the circumference of the circle, over its radius and over the interior points of the corresponding disc, respectively. It is therefore not possible to answer the question of which – if any – of the three models is the best model unless we specify precisely how the random experiment is going to be conducted.

Physical experiments

In 2002, a group of four first-year students at the Foundational Science Programme at Roskilde University wanted to make empirical experiments to see whether they could make physical realisations of Bertrand's thought experiment in order to distinguish and possibly choose between the three different idealised models (Skytte et al., 2002).

For this to be possible, they had various physical instruments made at the department workshop.

The first instrument was a quadratic metal plate on which the inscribed circle of the square and the inscribed equilateral triangle of the circle were engraved (see Figure 3.17). An arrow ("clock hand") whose length was the radius in the inscribed circle was fixed at one end at the centre of the circle so that it could rotate freely, eight to nine times before it stopped. One of the vertices of the inscribed triangle was chosen, and the arrow was rotated 6–8 times until it stopped, and it was observed whether the endpoint of the arrow was situated on the 120^0 arc between

72 Modelling examples

the opposite vertices or not. This experiment corresponds exactly to model 1. The experiment was repeated 100 times, and it was found that the arrow's endpoint was on this arc 33 times, corresponding to an empirical probability of 0.33, which was seen as providing confirmation that this physical realisation of the thought experiment is reasonably well captured by model 1.

In order to test model 2, other equipment was devised and produced (see Figure 3.18). First, a rectangular cardboard with dimensions 600 mm × 900 mm was

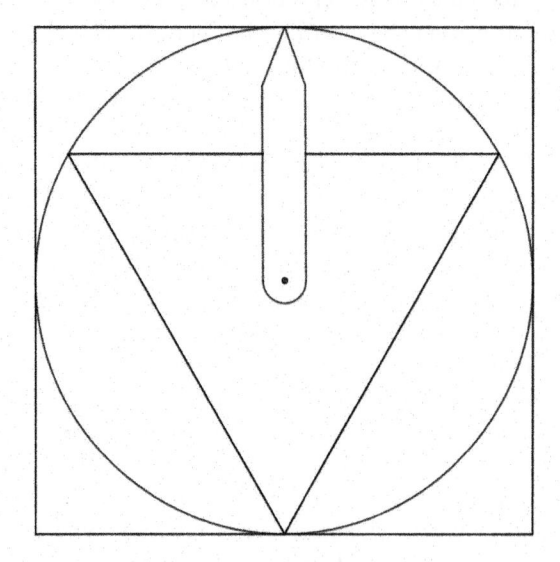

FIGURE 3.17 Metal plate for experiments concerning model 1

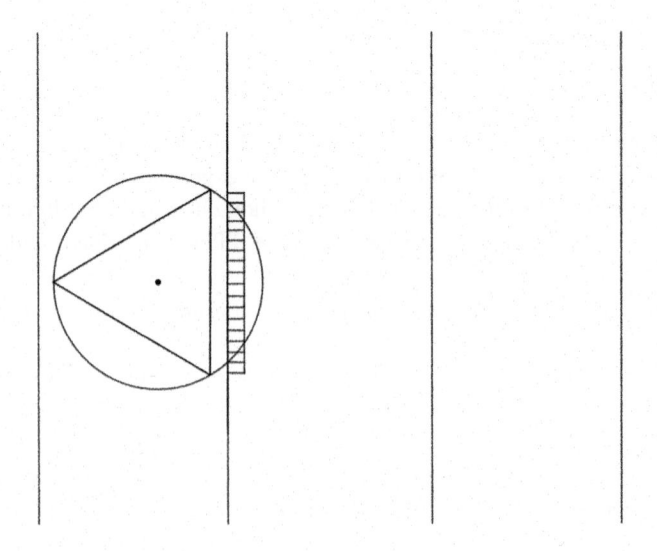

FIGURE 3.18 Cardboard with parallel lines for experiments concerning model 2

cut out and equipped with equidistant parallel lines, with a distance of 120 mm between the lines. Moreover, a flat planar disc was made of stainless steel, with a diameter of 120 mm. Also, a flat metal rod of length 103.9 mm ≈ $60\sqrt{3}$ mm, i.e., the length of the side of the equilateral triangle in the larger circle, was produced. It was assumed that throwing this disc from a suitably large distance onto the cardboard was a random experiment equivalent to throwing a chord at random on the circle with a diameter of 120 mm. The metal disc thrown onto the cardboard will exactly cut one of the parallel lines (except in the rare cases where two lines are exactly tangent to the outer circle). Using the metal rod, one can tell whether the length of the chord on the outer circle is larger than the length of the triangle side (see Figure 3.18, in which we have also drawn the inscribed equilateral triangle for illustration). This is equivalent to the chord lying closer to the centre of the circle than half the diameter, which is the core of model 2. To test the results of model 2, this experiment was performed 100 times. The outcome was that in 56 out of 100 throws, the chord was longer than the triangle side, i.e., the empirical probability was 0.56, which deviates by 12% from the model result of 0.5. Given that 100 throws are not that many, this result is not surprising. Once again, our physical experiment is a realisation of the thought experiment based on a uniform distribution of midpoints over a radius, as captured in model 2.

As finally regards model 3, a new piece of 600 mm × 900 mm cardboard was produced, this time equipped with a grid of two orthogonal bundles of equidistant parallel lines, with 120 mm as the distance between two neighbouring parallel lines (see Figure 3.19). Model 3 is based on the idea of randomly throwing a point (representing the midpoint of a chord) into the circle and checking whether it falls inside the inner circle of half the diameter. Instead of throwing the point into the circle, we can throw a metal ring with an outer diameter of 120 mm and an inner

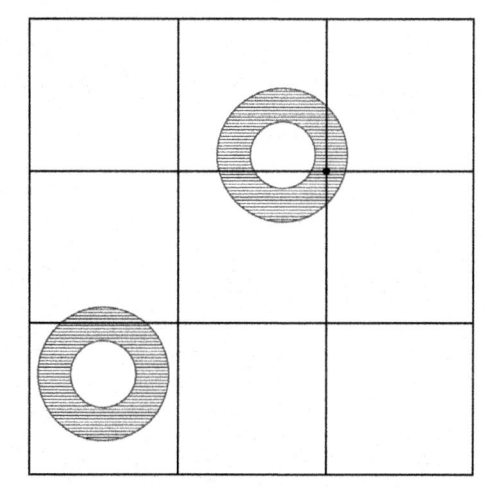

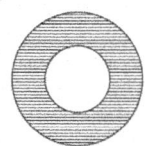

FIGURE 3.19 Metal ring and a cardboard with a grid for experiments concerning model 3

74 Modelling examples

diameter of 60 mm onto the plane and check whether a grid point lies in the inner circle or inside the metal ring between the two circles. While it will never happen that the entire ring/disc covers more than one grid point, it may happen, albeit seldomly, than no grid point is covered by the disc. In the rare cases where this actually did happen, the throw was not recorded. To cater for this, a total of 200 acceptable throws were made, and 49 of these gave rise to a grid point in the inner circle, corresponding to a chord longer than the side. The empirical probability is $49/200 \approx 0.245$, which is within 2% of what is predicted by model 3.

Concluding remarks

While one cannot, of course, consider the three experiments conducted by the students as scientific in any strict sense, they were sufficient to serve their purpose: to provide an empirical realisation of each of the three idealised models. In other words, the experiments demonstrate that all three idealised models do correspond to a realisable specification of what it may mean to throw a chord at random onto a circle. In more general terms, the thought experiment in combination with the physical experiments shed light on the very notion of randomness in the context of stochastic modelling and on the necessity of being extremely careful in specifying the nature of the random experiment lying behind any stochastic model. It is interesting to observe that, here, the realisations of three physical experiments were the judges in evaluating the theoretical models. Normally, the physical experiment comes first and a model is then constructed to capture it.

In conclusion, the physical experiments show that the apparent paradox in the models resulting from Bertrand's thought experiment disappears once the randomness involved in the three models is specified.

Notes

1 The values of all side lengths and widths are irrational, which implies that the decimal expansions are infinite and aperiodic. Any cut off after a final number of decimals will therefore be an approximation. Of course, beyond a few decimals, say two or three, this has no practical significance.
2 To show this, let v_0 be the value of v for which $1/v = sv$, i.e., $s = 1/v_0^2$. Then $g(v) = 1/v + v/v_0^2$, for $v > 0$, and $g(v_0) = 2/v_0^2$. Then we have $g(v) \geq g(v_0)$ if and only if $1/v + v/v_0^2 \geq 2/v_0$, which (by multiplication of both sides with the positive number vv_0^2) is equivalent to $v_0^2 + v^2 \geq 2vv_0$, which in turn is equivalent to $(v_0 - v)^2 \geq 0$. Since the latter inequality always holds, we conclude that indeed $g(v) \geq g(v_0)$ for any $v > 0$, which means that v_0 is the (uniquely determined) minimum point for g.
3 As a matter of fact, one might imagine several other idealised models of the thought experiment. For example, since the lengths of chords are all elements in the interval $[0,2r]$, the chords with length larger than the side of the inscribed equilateral triangle, i.e., $r\sqrt{3}$, have lengths in the sub-interval $[r\sqrt{3}, 2r]$. Assuming a uniform distribution of chord lengths over the interval $[0,2r]$, the probability that a chord is longer than the side is $(2 - \sqrt{3})/2 = 1 - \sqrt{3}/2$. Another possibility is to consider areas instead of lengths in models 1 and 2. In both models, all chords *shorter* than a side of the equilateral triangle constitute an area

consisting of two circle sections above one of its sides. The area of one of these sections is $1/3 \cdot (\pi - 3/4 \cdot \sqrt{3})r^2$, so the ratio of the areas of two sections and the whole circle is $2/3 - \sqrt{3}/2\pi$, hence the probability that the chord is longer than one side is $1/3 + \sqrt{3}/2\pi$.

References

Bertrand, J. (1889). *Calcul des Probabilités*. Paris: Gauthier-Villars.
Blum, W. & Leiß, D. (2005). Modellieren im Unterricht mit der "Tanken"-Aufgabe. In: *mathematik lehren* **128**, 18–21.
Kac, M. (1984). Marginalia: More on randomness. In: *American Scientist* **72**(3), 282–283.
Skytte, B.G., Baagøe, L.L., Lind, A.R. & Nørby, S.M. (2002). *Nåle, linier og cirkler – et projekt om tilfældighed (Needles, Lines and Circles: On Stochastic Phenomena)*. Roskilde: NatBas.
Tessier, P.E. (1984). Bertrand's paradox. In: *The Mathematical Gazette* **68**(443), 15–19.

4
MODELLING COMPETENCY AND MODELLING COMPETENCIES

4.1 Introduction and brief historical outline

The interest in the teaching and learning of mathematical applications and modelling that gained international momentum in the late 1970s soon expanded to include a focus on students' ability to engage in and undertake mathematical modelling processes. Researchers in the field – especially researchers belonging to the ICTMA community (see section 2.6) – began to speak of modelling achievement, modelling behaviour, modelling skills (Burkhardt, 1984; Galbraith & Clatworthy, 1990; Edwards & Hamson, 1996; Stillman, 1998; Haines et al., 2001) and modelling capabilities (Niss, 2001). Around the turn of the 21st century, the notion of modelling competency and competencies as the common term for these abilities gradually came into general use.

One of the first publications on modelling in mathematics education, as opposed to the study of applications and models, attracting international attention, was the set of five modelling units distributed in the UK by the Open University (1978). As observed by various researchers, including Maaß (2006, p. 117) and Kaiser and Brand (2015, p. 132), early notions of modelling skills, abilities and competencies originated in researchers' attempts to assess students' modelling work, typically in the context of group work in which activities finished with a written report explaining, justifying and accounting for the work done. Thus Galbraith and Clatworthy (1990, p. 145) specified four assessment criteria stated in terms of modelling abilities: ability to specify the problem clearly; ability to formulate an appropriate model: choose variables and find relations; ability to solve the mathematical problem including mathematical solution, interpretation, validation and evaluation/refinement; and ability to communicate results in a written and oral form. Similarly, Money and Stephens (1993) described the assessment criteria for a Common Assessment Task

Investigative Project, not just for some but for all upper secondary school students in Victoria, Australia, as follows:

> CAT1 is intended to enable students to demonstrate their ability to carry out an extended piece of independent work, define important variables, simplify complex situations, formulate useful questions and interpret problems mathematically, seek out and use available resources, synthesize and analyse information, and organize, structure, and communicate mathematical ideas and results.
>
> *(p. 327)*

Similar components were put forward by Ikeda & Stephens (1998, p. 227):

> (G1) Did the students identify the key mathematical focus of the problem? (G2) Were relevant variables correctly identified? (G3) Did the student idealize or simplify the conditions and assumptions? (G4) Did the student identify a principal variable to be analyzed? (G5) Did the student successfully analyze the principal variable and arrive at appropriate mathematical conclusions? (G6) Did the student interpret mathematical conclusions in terms of the situation being modelled?

Haines et al. (2001) set out (among others) to answer the question "How are modelling skills improved?" (p. 367) by putting all the stages of modelling – i.e., formulating model, solving mathematics, interpreting outcomes, evaluating solution, refining model, reporting – under a "microscope", as they wrote (p. 368). To that end, they developed a test instrument focusing on students' generating assumptions; clarifying questions; specifying problem statements; choosing relevant parameters, variables and constants; and choosing between different final models (pp. 377–379).

The first time the term "competency" was introduced into the context of mathematical models and modelling was in a master's thesis at Roskilde University in 1996 (Hansen et al., 1996), *Model Competencies – Developing and Testing a Conceptual Framework* (our translation from Danish), supervised by Morten Blomhøj. In this thesis, the authors decomposed what they evidently perceived as an aggregate mathematical model(ling) ability into a set of sub-competencies (structuring competency, mathematisation competency, de-mathematisation competency, validation competency, strategic competency, reflection competency, critical competency and communication competency).

In the late 1990s, the Danish KOM project (Niss & Jensen, 2002; Niss & Højgaard, 2011) and a number of derived or parallel projects – e.g., "Adding It Up" (Kilpatrick et al., 2001) and the RAND Mathematics Study Panel (2003) and, a little later, the German *Bildungsstandards* programme (Blum et al., 2006) – in different parts of the world placed emphasis on the active, general *enactment* of mathematics rather than solely on the knowledge of mathematics and related skills. In these projects, mathematical modelling is one in a set of several competencies that span the range of mathematical activities that students are meant to master and usually encounter in schools and universities. We describe one such set in some detail, such as it was construed in the Danish KOM project, since this project was the first one to deal with the question of mathematical competencies in a comprehensive manner.

In the KOM project, the focus is on the general notion of mathematical compete*nce* – i.e., an individual's capability and readiness to act appropriately and in a knowledge-based manner, in situations and contexts that involve actual or potential mathematical challenges of *any* kind. It went on to introduce eight mathematical compete*ncies*, each of which deals with situations and contexts involving a *particular* kind of mathematical challenge that are meant to jointly constitute mathematical competence at large. These competencies are: mathematical thinking competency, problem handling competency, modelling competency, reasoning competency, representation competency, symbolism and formalism competency, communication competency, and aids and tools competency. In the KOM project and related projects, the modelling competency is one of the key competencies constituting mathematical competence.

Based on Niss and Jensen (2002), and Niss and Højgaard (2011), modelling competency can be defined as follows: It consists of two interrelated components. The first component is the ability to actively *construct* mathematical models in various domains, contexts and situations, i.e., bringing mathematics into play in dealing with extra-mathematical matters. Active construction of models – oftentimes simply called "active" modelling – further involves a number of elements as described in Chapter 2: pre-mathematising the extra-mathematical context and situation to be modelled, including specifying the questions to be answered by the model; mathematising the situation to construct a mathematical model with reference to some mathematical domain; undertaking mathematical treatments (including problem solving) within that domain so as to obtain mathematical conclusions concerning the model at issue; de-mathematising the model by translating the conclusions obtained back into the extra-mathematical domain and interpreting them in terms of conclusions pertaining to the situation modelled; validating these conclusions, i.e., the answers provided by the model; and – finally – evaluating the entire model.

The other component is the ability to *analyse* given mathematical models, constructed by others or by oneself, and their foundation and to critically examine and assess their scope and validity. This includes "de-mathematising" (features of) such models, i.e., de-coding and interpreting model elements and results in relation to the context and situation being modelled as well as to the very aim of the model.

A natural question to ask here is: "What is the relationship between modelling competency and the modelling cycle?" It follows from the above definition that modelling competency encompasses the ability to carry out the full modelling cycle (in any version, see section 2.3) in a wide variety of contexts and situations. However, it also follows that modelling competency goes beyond that ability by also including the ability to understand and critically relate to, analyse and assess extant models.

The rapidly growing focus on modelling competency and modelling competencies is reflected in several publications from the first years of the 21st century. A significant example can be found in the Discussion Document for the 14th ICMI Study, "Modelling and Applications in Mathematics Education" (Blum et al., 2003), in which Issue 3a reads (p. 210):

> Issue 3a. How can modelling ability and modelling competency be characterised and how can it be developed over time?

and lists a number of specific questions, including:

> Can specific subskills and subcompetencies of "modelling competency" be identified?

It is therefore only natural that the resulting Study Volume (Blum et al., 2007) contains an entire part (3.3.) titled "Modelling Competencies" (pp. 217–264). The same is true of the book resulting from ICTMA-12, held in London in 2005 (Haines et al., 2007), where the section "Recognising Modelling Competencies" (pp. 90–175) was devoted to characterising, analysing and discussing modelling competency and competencies. This is also true of Part V (pp. 343–437) of the book coming out of ICTMA-14, held in Hamburg 2009 (Kaiser et al., 2011).

In other words, "modelling competency" and "modelling (sub-)competencies" seem to be here to stay, at least as far as the terminology is concerned. As Kaiser and Brand put it (2015, p. 135): "[I]t can be concluded that modelling competencies can be seen as a settled topic in the current modelling discussion, being promoted in various projects all over the world". In the sections to follow, we shall say more about what is actually covered by these and related terms.

4.2 Modelling competency/cies: cognition and volition

It is a general issue in the discussion of mathematical competencies, including the modelling competency/cies, whether volitional (i.e., attitudinal, dispositional and affective) aspects should be included in the notion of competency or whether these aspects should be perceived as separate, leaving the competency notion to be defined as a purely cognitive one. Since a proposed definition can never be true or false, only more or less appropriate and useful, this is not an issue of what is correct and incorrect – both positions are easily defendable – but an issue of the purposes of the discussion as well as of terminological transparency. While everyone acknowledges the importance of volitional issues, the researchers associated with the KOM project have chosen to keep the notion of competency as a purely cognitive one and to consider the volitional aspects as belonging to a conceptually different category. In contrast, German researchers, inspired by the psychologist Weinert (2001), have chosen to include volitional aspects as an integral part of the definition of mathematical competencies in general and of modelling competency in particular. Thus Kaiser (2007, p. 110) reserves the term "ability" for the purely cognitive notion and defines "competency" as including volitional elements: "Modelling competencies include, in contrast to modelling abilities, not only the ability but also the willingness to work out problems, with mathematical aspects taken from reality, through mathematical modelling." This is also the point of view of the abovementioned German education standards which explicitly refer to Weinert. In this book, we take "competency" to have a purely cognitive meaning unless explicitly stated differently.

4.3 Modelling competency and modelling (sub-) competencies

If one examines the literature on modelling competency and competencies, it soon becomes clear that, from a conceptual point of view, different researchers have adopted two rather different perspectives on and definitions of these notions.

The first definition, which we shall term the *"top-down" definition*, deals with a comprehensive, overarching entity called *the* modelling competency in the singular. According to this view, there exists such a distinct, recognisable and more or less well-defined entity. It is typically possible, as a result of a closer analysis, to identify major components of and other elements in this entity, but *the* modelling competency nevertheless is the primary object, whereas the major components – oftentimes named sub-competencies – are derived, secondary objects.

The second definition, the *"bottom-up" definition*, deals with a set of distinct and separate modelling competen*cies* in the plural without, in the first place, seeing them as instances, aspects or components of a comprehensive, overarching modelling competency. These competencies are tightly linked to the modelling cycle.

At first glance, it may seem to be mainly a fight about words and terminology whether we consider a comprehensive, overarching modelling competency and then decompose it into a set of sub-competencies (as in the top-down definition) or whether we take our point of departure in a set of separate, independent modelling competencies and – perhaps but not necessarily – subsequently aggregate these into a composite modelling competency, of which the initial competencies then might be perceived as sub-competencies (the bottom-up definition). However, there is much more at stake than words and terms. First, the two views and definitions are conceptually and empirically different. In the top-down view, the modelling competency is the primary complex notion which can be subjected to conceptual, theoretical and empirical analysis, thus giving rise to a multitude of different components, elements and prerequisites, depending on the perspective taken. This notion corresponds to an empirically well-delineated entity found in the real world: the ability to construct and analyse models of extra-mathematical domains and situations. We can say that someone possesses this competency if s/he is able to deal successfully with such challenges. In the bottom-up view, a set of different, separate modelling competencies constitute the primary notions, each of which exists – in conceptual and empirical terms – independently of the other competencies under consideration. Someone may possess some of these competencies and not others, depending on his/her ability to deal with particular aspects of modelling processes. Only when bundled together can they give rise to an aggregate, overarching notion of modelling competency. The top-down definition acknowledges that there is more to modelling competency than the set of sub-competencies it entails. The bottom-up view cannot admit that the "whole is more than the sum of its parts" since modelling competency, if considered at all, is nothing but the definitional aggregation of the sub-competencies it is composed of.

This distinction has didactical and pedagogical consequences as well. The top-down view lends itself to both a holistic (see Blomhøj & Jensen, 2003, and also Stillman, 1998) educational approach to the modelling process in its entirety, corresponding to teaching and learning activities in modelling encompassing the full modelling cycle, and to an atomistic approach to some of the sub-competencies, corresponding to teaching-learning activities concentrating on one or a few components of the modelling process at a time. In contrast, the bottom-up view only lends itself to an atomistic approach to the various modelling competencies. To be sure, even if students are working in settings involving the full cycle, the focus on the individual sub-competencies remains atomistic in nature. As stated by Kaiser & Brand (2015, p. 138): "The overall achievement is based on the achievement within different sub-competencies." It is important, however, to be aware that the distinction between the top-down and bottom-up definitions is of a conceptual nature, whereas the distinction between a holistic and an atomistic approach to modelling work is of a didactico-pedagogical nature, where the generic question is "What is the optimal blend of holistic and atomistic modelling work so as to foster and promote modelling competency?"

It should be noticed that the distinction between the top-down and the bottom-up definitions of modelling competency, respectively competencies, has been uncovered by analysing the expositions in several publications. With a few exceptions – e.g., Kaiser and Brand (2015, p. 139), who write "How do we describe and conceptualise modelling competency, as an underlying general ability or composed of different sub-abilities?" – authors typically have not made this distinction, at least not explicitly, which suggests that they may not have noticed it. It is, however, possible that researchers who have adopted a bottom-up definition tacitly take it for granted that the modelling competencies they specify and consider do refer to an overarching yet un-named comprehensive modelling competency. Whether or not this is the case in each context has to be determined by undertaking a specific interpretative analysis of the particular context.

As is evident from the quotes above, the top-down view and definition are explicitly adopted by the KOM project (Niss & Jensen, 2002; Niss & Højgaard, 2011). This is also the view adopted by Blomhøj and Jensen (2003) and by Jensen (2007). Kaiser-Meßmer, in her PhD dissertation (1986), seems to have taken this perspective as one of the first researchers in the field.

The bottom-up definition appears to be the one chosen in an unpublished fund application paper by Blum and Kaiser (1997), quoted in Maaß (2006, pp. 116–117). The paper speaks about five sets of application competencies (in German "Anwendungsfähigkeiten"), each of which was divided into several sub-competencies (without using that term): competencies to understand the real problem and to set up a model based on reality; competencies to set up a mathematical model from the real model; competencies to solve mathematical questions within this mathematical model; competencies to interpret mathematical results in a real situation; and competencies to validate the solution. Blum and Kaiser summarise this as follows: "Global gesehen verstehen wir unter mathematischen Anwendungsfähigkeiten die Fähigkeit von Lernenden, bekannte oder selbst zu entwickelnde mathematische Methoden zum Beschreiben, zum besseren Verstehen und zum Bewältigen von außermathematischen Situationen zu verwenden" (quoted in Maaß, 2004, p. 32). ("Globally considered, by mathematical application competencies we understand the learner's ability to employ established or personally created mathematical methods and representations to better understand and master extra-mathematical situations", our translation). This should clearly be interpreted as an aggregation of a broad variety of independent application (modelling) competencies into one over-arching application (modelling) competence.

Maaß (2006) offers a survey of parts of the conceptual development of all these notions. Based on the aforementioned list of modelling competencies, which she seems to perceive as a list of separate competencies, Maaß sets out to empirically look for these and other possible competencies in lower secondary students' work on a series of modelling tasks, thus also adopting a bottom-up approach. She finds that all these competencies actually exist as significant prerequisites for successful modelling, but they need to be supplemented by other competencies, namely:

> Metacognitive modelling competencies; Competencies to structure real world problems and to work with a sense of direction for a solution; Competencies to argue in relation to the modelling process and to write down this argumentation; Competencies to see the possibilities mathematics offers for the solution of real world problems and to regard these possibilities as positive.
>
> *(p. 139)*

One might perceive these additional competencies as aspects of one aggregate, over-arching modelling competency, even though Maaß does not phrase her conclusion in this way. Instead, she states her conclusion as follows: "[M]odelling competencies include more competencies than just running through the steps of a modelling process" (p. 139). She observes that different researchers' conceptualisations of modelling competencies are intimately related to how the modelling process is being understood and conceptualised and hence to the representation of this process in terms of some version of the modelling cycle. In other words, for these researchers modelling competencies seem to be basically defined in terms of the modelling cycle.

Kaiser and Brand (2015), after having stated in the beginning of their article (p. 129) that there still doesn't exist a common understanding of modelling competencies among researchers, identified four major strands in the debate on modelling competencies (p. 135):

- Modelling competencies in an overall comprehensive concept of mathematical competencies (as in the Danish KOM Project);
- The assessment of modelling skills and the development of assessment instruments (as with certain Australian and British researchers);
- The development of a comprehensive concept of modelling competencies based on sub-competencies (predominant among German researchers); and
- The integration of meta-cognition into modelling competencies (predominant among Australian and Portuguese researchers and others).

The first of these strands corresponds to the top-down view mentioned above, according to which the primary notion is modelling competency rather than competencies, whereas the second and third strand correspond to the bottom-up view. The fourth strand doesn't represent a specific stance in relation to the two views but is placed somewhere between them. It is interesting to note that in the concluding paragraph of their paper, where the authors report on an own empirical investigation into the relationship between holistic and atomistic approaches to teaching and learning of modelling competency, Kaiser and Brand (2015) write (p. 146) "[T]he study suggests that this complex construct [modelling competency] can be described as a global overarching modelling competency and several sub-competencies", which is a way of combining the top-down and the bottom-up definitions.

4.4 Holistic and atomistic approaches to the development of modelling competency

Above, we touched upon the notions of holistic and atomistic approaches to modelling competencies. Even though Stillman (1998, pp. 244–245) speaks about a holistic view of the messy, complex modelling process, the distinction was first explicitly made by Blomhøj and Jensen (2003). It was not made to distinguish between different definitions of modelling competency, respectively (sub-)competencies (they both work from the top-down definition). Rather, it was to distinguish between different ways of orchestrating teaching and learning activities in order to foster and develop the modelling competencies in students. In the holistic approach, students work on the modelling process in its entirety, dealing with all relevant extra-mathematical and mathematical components in the same context, including all those represented in the modelling cycle. In the atomistic approach, students work to develop one or a few (sub-)competencies at a time. Blomhøj and Jensen (2003) mention mathematising and analysing models as examples (p. 128), typically by dealing with several tasks, each of which involves one or two stages in the modelling cycle only. By moving around in the modelling cycle, it is expected that students accumulate experiences with and hence develop, by way of aggregation, the whole set of modelling (sub-)competencies.

The obvious but simplistic question then is: Which of the two approaches, the holistic or the atomistic one, is the more effective for students to develop modelling competency and (sub-)competencies? In case this question doesn't have a definitive and clear-cut answer (which it most probably doesn't), a less simplistic, natural follow-up question is: What is an efficient balance between the two approaches, and how can it be achieved in practical terms? These are empirical didactico-pedagogical questions, pertaining to both research and practice.

In a research context, it is important to realise that it is not meaningful to compare the effectiveness of the two approaches unless there is a common conceptual basis for the comparison. This requires a common definition of modelling competency and (sub-)competencies. If the holistic approach were carried out with reference to the top-down definition of mathematical competency and sub-competencies, whereas the atomistic approach were carried out with reference to a version of the bottom-up definition that denies the existence of a comprehensive overarching modelling competency, a complete comparison of the two approaches would hardly make sense since the criteria for success of the respective approaches would not be the same. It would only be possible to make a partial comparison of the two approaches consisting of a comparison of their respective abilities to foster and develop students' different sub-competencies of the modelling competency (in the bottom-up definition). If, however, the version of the bottom-up definition is one which is accompanied by a subsequent aggregation of the separate modelling competencies into a global overarching modelling competency, as in Kaiser and Brand (2015, p. 146), there does exist a common ground for a comparison between the two approaches.

Relatively little research has been done to systematically answer the questions just stated. Research and research findings are the subject of Chapter 6. Suffice it here to say that Zöttl et al. (2011) have conducted a study which concludes that assessment based on the atomistic approach (which they call the analytic approach) is superior to assessment based on the holistic approach. Kaiser and Brand (2015), in contrast, report a comparative study of a "holistically taught" and an "atomistically taught" group of students which found that, in some respects, the holistic approach outperformed the atomistic one, whereas in other respects the converse was the case. This certainly suggests that these issues are far from being settled.

4.5 Modelling competency and other competencies

As mentioned in section 4.1, in the Danish KOM-project the modelling competency was one of eight competencies, cf. the so-called competency flower in Figure 4.1.

The KOM competency flower shows that each competency has a distinct identity of its own but that, nevertheless, all competencies overlap. This means that the activation of each competency in each context typically draws upon several other competencies. It depends on the context which competencies are actually drawn upon and how.

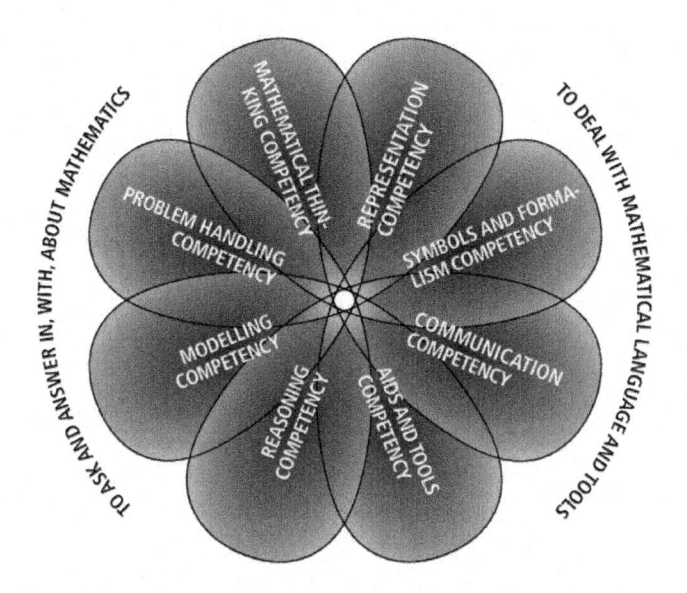

FIGURE 4.1 The KOM competency flower

Successful activation of the modelling competency usually involves, at least, the problem handling, representation, symbols and formalism, reasoning, and communication competencies and oftentimes the aids and tools competency and the thinking competency as well.

Thus, with the taxi example addressed in section 2.2, the thinking competency is in play when the modeller considers what kinds of mathematical questions might be relevant in the situation. The problem handling competency is involved when the modeller settles on a problem-solving strategy consisting of deciding first to specify the parameters of the linear function at issue, then to identify the mathematical questions that have to be answered with regard to this function, and finally how the answers should be obtained. The symbols and formalism competency is significant in specifying the places and roles of parameters and the variables of the function considered and in subsequently performing the calculations needed to determine the relevant value of this function, whereas the representation competency does not need to be activated beyond the algebraic representation of the linear function, already included in the aspects of the symbols and formalism just mentioned. The aids and tools competency may be activated by someone who is unable or unwilling to perform the calculations needed to find the relevant function value. The reasoning competency is not invoked here beyond ascertaining that the conclusion obtained in determining the function value is correct with respect to the computation algorithms used. A modeller who does not make the calculations to find the function value can only justify the answers by repeated use of technology, possibly through different media. Finally, the involvement of the communication competency depends on the communicative setting in which the modeller is placed. To a student who is showing and defending her or his work to peers or to the teacher, the communication competency is invoked in a rather complex manner where the disentanglement from one another of the real-world components, the modelling components and the purely mathematical components are key points.

It is worth emphasising that since the enactment and exercising of the modelling competency also activates and draws upon the other competencies, the modelling competency serves to enhance and consolidate aspects of these other competencies. However, it would be wrong to conclude that the modelling competence alone would then suffice for the other competencies to be developed and exercised, for not all aspects of these competencies are invoked and activated by enacting and exercising the modelling competency.

Conversely, it might be tempting to think that developing the other seven competencies would automatically give rise to the development of the modelling competency as well. However, this is not true because it is well known from research (see Chapter 6) that one can possess a high level of intra-mathematical competencies without possessing the modelling competency, which in fundamental ways requires extra-mathematical domains be paid serious attention with mathematical modelling as the bridge linking these domains to mathematics.

4.6 Dimensions of possessing modelling competency

The KOM project (Niss & Jensen, 2002; Niss & Højgaard, 2011) put forward three dimensions of an individual's possession of any given mathematical competency: degree of coverage, radius of action, and technical level.

The *degree of coverage* of a competency is the extent to which all the aspects that define and characterise it form part of the individual's possession of the competency. As to the modelling competency, the degree of coverage for an individual who can both construct mathematical models and analyse extant models is higher than the degree of coverage for an individual who is "only" able to analyse already-given models. Also, the degree of coverage for an individual who can successfully manage all the stages and steps of the modelling cycle is higher than for an individual who can only manage a proper subset of these stages and steps, e.g., the mathematical treatment and de-mathematisation.

The *radius of action* in someone's possession of a competency is the range and variety of different contexts and situations in which the individual can successfully activate that competency. With regard to the modelling competency, an individual who can, e.g., activate it in contexts and situations pertaining to everyday household practices, including home economics, as well as in contexts and situations involving carpentry or growth phenomena in economics, financing and human populations, has a larger radius of action in his/her modelling competency than does someone who can only handle situations arising in everyday household practices.

The *technical level* of an individual's possession of a competency denotes the level and degree of sophistication of the mathematical concepts, results, theories, methods and techniques which the individual can bring to bear when exercising the competency. Thus, an individual who can, e.g., successfully handle modelling contexts and situations implicating linear, power, polynomial, exponential and trigonometric functions and their properties, as well as theoretical or empirical probability distributions, possesses the modelling competency at a higher technical level than is the case for an individual who can only successfully handle contexts and situations implicating linear functions and elementary descriptive statistics.

None of the three dimensions just outlined constitutes a complete ordering, let alone a full ranking, of the corresponding aspects of competency possession. Rather, the ordering of competency possession with respect to each of the dimensions is a partial and qualitative one. This fact becomes even more pronounced when several dimensions are considered at the same time. Thus, an individual's possession of the modelling competency may have a high degree of coverage in everyday household practices, for example, with respect to managing the stages and steps of the modelling cycle, but a low degree of coverage in contexts and situations from science or technology. Similarly, another individual may possess the modelling competency at a large radius of action but at a low technical level.

Despite these caveats, the three dimensions serve the purpose of articulating, in qualitative terms, significant aspects of competency possession and variability. They also may serve as an analytic instrument to uncover the range and scope of what we might call a generic modelling competency across several mathematical domains and across different extra-mathematical contexts and situations. In other words, they can help address the question: "To what extent is modelling situated?" From an experiential point of view, there does exist a context-free modelling competency – otherwise, we wouldn't be teaching general modelling. On the other hand, the liberation from contexts is not and cannot be without limits. So what are these limits? This is a very pertinent research question to which we shall return in Chapter 6.

4.7 Final remarks

The exposition above shows that there is universal agreement among researchers in the field that the notions of modelling competency and modelling (sub-)competencies are key constructs in research and practice in mathematical modelling. However, it also shows that there is not yet agreement on how these notions are to be defined. Two issues remain unsettled: First, the more general issue – pertaining to mathematical competencies in general and to the modelling competency/cies in particular – concerns whether competencies should be defined as purely cognitive constructs or whether the definitions should explicitly include volitional elements. The second addresses the specific issue of whether modelling competency and modelling (sub-)competencies should be given a top-down definition in which an overarching comprehensive modelling competency is considered the primary notion whereas sub-competencies are secondary, derived notions; whether a bottom-up definition based on a set of separate modelling competencies, perhaps subsequently bundled together to form a global modelling competency, is to be preferred; or whether a compromise between the two definitions should be made.

Finally, the empirical question of how the balance should be struck between holistic and atomistic approaches to developing modelling competency/cies in students is far from settled. However, any meaningful comparison between the two approaches presupposes a common conceptual ground to be found for the definition of modelling competency/cies.

One should keep in mind, however, that different viewpoints on a *problématique* can also be considered as complementary rather than antagonistic, opening avenues for deeper understanding of the *problématique*. Choosing between different definitions and approaches may well be a matter of purpose, practical use or taste, but it remains important to avoid that conceptions which are different in essence are mixed up.

References

Blomhøj, M. & Jensen, T.H. (2003). Developing mathematical modelling competence: Conceptual clarification and educational planning. In: *Teaching Mathematics and Its Applications* **22**(3), 123–139.

Blum, W., Alsina, C., Biembengut, M.S., Confrey, J., Galbraith, P., Ikeda, T. et al. (2003). Discussion document for the fourteenth ICMI study applications and modelling in mathematics education. In: *L'Enseignement mathématique* **49**(1–2), 205–214.

Blum, W., Drüke-Noe, C., Hartung, R. & Köller, O. (Eds.) (2006). *Bildungsstandards Mathematik: konkret*. Berlin: Cornelsen.

Blum, W., Galbraith, P.L., Henn, H.-W. & Niss, M. (Eds.) (2007). *Modelling and Applications in Mathematics Education: The 14th ICMI Study*. New York, NY: Springer.

Blum, W. & Kaiser, G. (1997). Vergleichende empirische Untersuchungen zu mathematischen Anwendungsfähigkeiten von englischen und deutschen Lernenden. *Unpublished application to Deutsche Forschungsgemeinschaft*.

Burkhardt, H. (1984). Modelling in the classroom: How can we get it to happen? In: J.S. Berry, D.N. Burghes, I.D. Huntley, D.J.G. James & A.O. Moscardini (Eds.), *Teaching and Applying Mathematical Modelling*. Chichester: Horwood.

Edwards, D. & Hamson, M. (1996). *Mathematical Modelling Skills*. London: Macmillan.

Galbraith, P.L. & Clatworthy, N.J. (1990). Beyond standard models: Meeting the challenge of modelling. In: *Educational Studies in Mathematics* **21**(2), 137–163.

Haines, C.R., Crouch, R. & Davies, J. (2001). Understanding students' modelling skills. In: J.F. Matos, W. Blum, S.K. Houston & S.P. Carreira (Eds.), *Modelling and Mathematics Education: ICTMA 9: Applications in Science and Technology* (pp. 366–380). Chichester: Horwood.

Haines, C.R., Galbraith, P.L., Blum, W. & Khan, S. (Eds.) (2007). *Mathematical Modelling: ICTMA 12: Education, Engineering and Economics*. Chichester: Horwood.

Hansen, N.S., Iversen, C. & Troels-Smith, K. (1996). *Modelkompetencer – udvikling og afprøvning af et begrebsapparat*. Tekster fra IMFUFA, No. 321. Roskilde: IMFUFA, Roskilde Universitet.

Ikeda, T. & Stephens, M. (1998). The influence of problem format on students' approaches to mathematical modelling. In: P.L. Galbraith, W. Blum, G. Booker & I. Huntley (Eds.), *Mathematical Modelling, Teaching and Assessment in a Technology-Rich World* (pp. 223–232). Chichester: Horwood.

Jensen, T.H. (2007). Assessing mathematical modelling competency. In: C.R. Haines, P.L. Galbraith, W. Blum & S. Khan (Eds.), *Mathematical Modelling (ICTMA 12), Education, Engineering and Economics* (pp. 141–148). Chichester: Horwood.

Kaiser, G. (2007). Modelling and modelling competencies in school. In: C.R. Haines, P. Galbraith, W. Blum & S. Khan (Eds.). *Mathematical Modelling (ICTMA 12): Education, Engineering and Economics* (pp. 168–175). Chichester: Horwood.

Kaiser, G., Blum, W., Borromeo Ferri, R. & Stillman, G. (Eds.) (2011). *Trends in the Teaching and Learning of Mathematical Modelling.* Dordrecht: Springer.

Kaiser, G. & Brand, S. (2015). Modelling competencies: Past development and further perspectives. In: G. Stillman, W. Blum & M.S. Biembengut (Eds.), *Mathematical Modelling in Education Research and Practice: Cultural, Social and Cognitive Influences* (pp. 129–149). Cham, Heidelberg, New York, Dordrecht, London: Springer.

Kaiser-Meßmer, G. (1986). *Anwendungen im Mathematikunterricht* (Empirische Untersuchingen, Vol. 2). Bad Salzdetfurth: Franzbecker.

Kilpatrick, J., Swafford, J. & Findell, B. (Eds.) (2001). *Adding It Up: Helping Children Learn Mathematics.* Mathematics Learning Study Committee. Center for Education, Division of Behavioral and Social Sciences and Education. Washington, DC: National Academy Press.

Maaß, K. (2004). *Mathematisches Modellieren im Unterricht – Ergebnisse einer empirischen Studie.* Hildesheim: Franzbecker.

Maaß, K. (2006). What are modelling competencies? In: *ZDM: The International Journal on Mathematics Education* **38**(2), 113–142.

Money, R. & Stephens, M. (1993). Linking applications, modelling and assessment. In: J. de Lange, I. Huntley, C. Keitel & M. Niss (Eds.), *Innovation in Mathematics Education by Modelling and Applications* (pp. 323–336). Chichester: Horwood.

Niss, M. (2001). Issues and problems of research on the teaching and learning of applications and modelling. In: J.F. Matos, W. Blum, S.K. Houston & S.P. Carreira (Eds.), *Modelling and Mathematics Education: ICTMA 9: Applications in Science and Technology* (pp. 72–88). Chichester: Horwood.

Niss, M. & Højgaard, T. (Eds.) (2011). *Competencies and Mathematical Learning: Ideas and Inspiration for the Development of Mathematical Teaching and Learning in Denmark.* Roskilde University: IMFUFA. English translation of Danish original (2002).

Niss, M. & Jensen, T.H. (2002). *Kompetencer og matematiklæring. Ideer og inspiration til udvikling af matematikundervisning i Danmark.* Uddannelsesstyrelsens temahæfteserie nr. 18. Copenhagen: The Ministry of Education.

The Open University (1978). *Mathematics Foundation Course, Mathematical Modelling Units 1 to 5.* Milton Keynes: Open University Press.

RAND Mathematics Study Panel (2003). *Mathematical Proficiency for All Students: Toward a Strategic Research and Development Program in Mathematics Education.* Santa Monica, CA: RAND.

Stillman, G. (1998). The Emperor's new clothes? Teaching and assessment of mathematical applications at the senior secondary level. In: P. Galbraith, W. Blum, G. Booker & I. Huntley (Eds.), *Mathematical Modelling, Teaching and Assessment in a Technology-Rich World* (pp. 243–253). Chichester: Horwood.

Weinert, F. (2001). Concept of competence: A conceptual clarification. In: D. Rychen & L. Salgnik (Eds.), *Defining and Selecting Key Competencies* (pp. 45–66). Göttingen: Hogrefe & Huber.

Zöttl, L., Ufer, S. & Reiss, K. (2011). Assessing modelling competencies using a multidimensional IRT approach. In: G. Kaiser, W. Blum, R. Borromeo Ferri & G. Stillman. (Eds.), *Trends in the Teaching and Learning of Mathematical Modelling* (pp. 427–437). Dordrecht: Springer.

5
CHALLENGES FOR THE IMPLEMENTATION OF MATHEMATICAL MODELLING

5.1 Global challenges

There are many justifications for the inclusion of modelling and real-world applications into mathematics teaching, beginning in first grade, as we pointed out in section 2.8. The question is not if and why modelling ought to be implemented but how this implementation can be realised so that the aims of mathematics teaching in general and the aims of dealing with modelling contexts in particular will be achieved. However, there are several challenges and sometimes even barriers that inhibit the implementation of modelling in education (see, e.g., Freudenthal, 1973; Pollak, 1979; De Lange, 1987; Blum & Niss, 1991; Burkhardt, 2004; Ikeda, 2007; Schmidt, 2009; Blum, 2015).

Before considering several more specific challenges, we call attention to three challenges of an overarching nature. The first challenge stems from the fact that the use and role of mathematics in societal, cultural, technical and scientific contexts and situations is largely invisible to non-experts (Niss, 1994, speaks of the "relevance paradox" to indicate the fact that while mathematics is extremely relevant in the world, many people find it irrelevant for their own lives). In a multitude of cases, mathematics is hidden behind the immediately accessible surface of objects, constructs, artefacts and systems that rely on mathematics in crucial ways. Examples include: insurance premiums; determining the outcomes of parliamentary voting schemes; traffic light control; tempered tuning of musical instruments; mathematical control theory employed by automatic pilots in aeroplanes and ships; error-correcting codes in the design of CDs; and predictions of the future development of the global average temperature. To the non-expert, these examples appear to be something that has to do with: insurance companies; elections and election authorities; electrical engineering technicians; musicians and specialists in musical instruments; aeronautical engineers, pilots or ship captains; producers of CDs; and people preoccupied with global warming – and mathematics seems to have nothing to do with all of this. When the place and role of mathematics in such extra-mathematical domains are invisible, it is a great challenge for students to see the need for mathematical modelling and an equally great challenge for teachers to support students in recognising hidden mathematics.

The second overarching challenge is the fact that while a solid knowledge and understanding of mathematics and general mathematical competence are necessary prerequisites for successfully engaging in mathematical modelling, we know from research (see Chapter 6) that these prerequisites are not sufficient for successful modelling. Students must learn modelling in large part by actually doing modelling, which also means that organisational frameworks, conditions and resources must be in place to foster and promote students' modelling activities. This aspect constitutes the third global challenge, present in many places: The frameworks and boundary conditions for mathematics teaching and learning, such as curricula, the organisation of school instruction, and the modes of summative assessment, imply that implementing modelling into everyday classrooms requires a systemic change in most countries. Discussing these issues Burkhardt (2018, p. 74) writes:

> In many countries education is a "hot" political issue with school system leadership making decisions of a technical kind that they would not contemplate in, for example, medicine. So we must recognize that politicians and other policy makers are part of the system and take their priorities into account if we are to develop models of change that actually improve teaching and learning . . . .
>
> . . . [I]n countries with "high-stakes" assessment the range of performance types that are assessed ensures that these performances are developed in the classroom. . . . In particular, if modelling is to happen in most classrooms this needs to be assessed in the tests. Yet changes to these tests are always a sensitive issue, with teachers understandably preferring the known to the unknown . . .
>
> Policy makers tend to attempt comprehensive reform – a new national curriculum, for example – which either is largely cosmetic or, if ambitious, places new demands on teachers and other professionals that are not matched with the support needed for them to meet those demands. . . . The most successful improvement models in our experience are based on gradual change – an approach taken for granted in medicine, of course.

In addition to these three overarching challenges, there is a multitude of more specific challenges to the implementation of mathematical modelling referring to:

1   The competencies, beliefs, motivations and attitudes of the students (section 5.2),
2   The competencies, beliefs, motivations and attitudes of the teachers (section 5.3),
3   The norms and rules for communication and interaction in the classroom (section 5.4),

4   The necessities imposed by assessment, especially summative assessment (section 5.5),
5   The materials available (section 5.6), and
6   The presence and role of digital tools (section 5.7).

When we mention such challenges, we certainly do not want to blame students or teachers for being barriers to the implementation of mathematical modelling in teaching and learning of mathematics. We only want to call attention to and discuss various relevant factors pertaining to such implementation. Most of these challenges have their roots in the properties and conditions characteristic of the education system of any given country, which itself is embedded in the culture of the part of the world to which the country belongs. For instance, the ways in which communication and interaction in the classroom take place are strongly influenced by cultural factors. This has obvious consequences for the possible implementation of modelling. For instance, as already mentioned, modelling competency can only be developed if learners perform modelling activities themselves, which requires an activity-based teaching methodology and ensuing classroom communication and interaction that may be in conflict with teaching, communication and interaction traditions that have developed over decades, if not centuries.

One consequence of such challenges and barriers is that in most countries around the world, there is a considerable gap between, on the one hand, the state-of-the-art of modelling in the research and development literature and, on the other hand, the actual implementation of modelling in curricula, institutions, classrooms, lessons and summative and formative assessment schemes. We will discuss some of these challenges in the following sections.

5.2   Challenges on the students' side

A look at modelling tasks (see the examples in Chapter 3) from a cognitive point of view clearly shows that for successfully dealing with such tasks, students have to possess both mathematical knowledge – which is necessary in the mathematisation phase as well as in the phase of working mathematically – and a fair amount of extra-mathematical knowledge, which is necessary when dealing with the given context and situation in a knowledgeable manner, especially when undertaking pre-mathematisation, de-mathematisation, validation and evaluation (see the description of these processes in Chapter 2). To perform the steps in the modelling process, the students also have to possess certain competencies such as understanding a given problem situation or task (which, in addition to general language proficiency, involves aspects of the mathematical communication competency; see section 4.5), translating specific extra-mathematical concepts and relations into mathematical entities, or interpreting mathematical objects in extra-mathematical contexts (both of which are sub-competencies of the modelling competency; see Chapter 4). These competencies are needed in a concretised form in relation to the given task, to its extra-mathematical context and to the mathematics involved, not (only) in a somewhat general form (see the discussion of the range of competencies in section 4.6). However, this presupposes either that these competencies have already been acquired in similar contexts or that the students are able to transfer their competencies from a known context to a new context – generally a high demand (for more details, see the discussion in section 6.1). For such transfer to happen, it is significant that students possess meta-cognitive knowledge and strategies (see section 6.4). When dealing with modelling tasks and problems, every step in the process may be a barrier to students (for details, see section 6.2). Altogether, with the inclusion of extra-mathematical examples, cases and tasks, mathematics lessons become more open(-ended), more demanding and less predictable for learners. A focus in mathematics lessons on concepts, rules and algorithms, as is often encountered in ordinary classrooms, may give students more security about the expected demands of lessons and especially of examinations and of assessment at large. Teachers need to take students' concerns about the unpredictable openness of modelling environments in lessons or modelling tasks in examinations seriously and engage in conversations with students in attempts to justify why such examples and tasks are part of the agenda, and handle students' apprehension by specifying the demands of the corresponding activities in a clear manner. This inevitably implies re-negotiating the usual didactical contract with the students (Brousseau, 1997; see section 5.4).

Another challenge on the students' side comes from what Treilibs et al. (1980) call the "few years gap problem". If a modelling task is included in a set of exercises referring to a specific mathematical topic area, students think they know what is expected of them ("It must be Pythagoras!") and don't actually try to identify appropriate mathematical tools for modelling the given situation. A modelling task will be taken more seriously if it is not clear from the outset what mathematics can or should be used. Students then often tend not to use appropriate tools that are in fact available to them (e.g., algebra in the "filling up" task). Sometimes, they fall back to draw on far too elementary tools or do not attempt to tackle the task at all (see also section 6.2). Such behaviour may result from the fact that if a mathematical topic has been treated recently, it may not yet be sufficiently consolidated and cross-linked with other mathematics to be used in a modelling context. It may take a long time, and several experiences with such uses, before students can independently apply the mathematics they have learned to new and open modelling tasks. Hence, when it comes to selecting or designing tasks for a given educational setting, there may be a bigger temporal distance (a "few years gap") to the time when the mathematics needed for the tasks was treated in the curriculum.

In addition to possessing knowledge, competencies and strategies, students must be willing to deal with extra-mathematical tasks and problems. This has to do with their picture of mathematics as well. Can it be part of the mathematics classroom discourse to deal with the amortisation of loans (see example 5 in Chapter 3), or is this part of social science or of everyday life outside school? Can the question of how fast one should drive in dense traffic (see example 6 in Chapter 3) be a topic in a mathematics lesson, or should it only be part of the training programme in a driving school? Such questions touch on the students' picture of mathematics – does dealing with extra-mathematical problems really belong to "mathematics"? For many students, the answer is "no". They have been socialised in school to regard mathematics as a collection of concepts, rules, algorithms, theorems and theories that have nothing to do with the world outside; hence, they see no point and no meaning in dealing with real-world problems in mathematics. Perhaps they have encountered word problems that have been conceived as a means for dressing up the mathematical substance in the words of some other field or discipline (see section 2.9), which does not, however, have to be taken seriously as pertaining to the real world. One might expect or hope that other school subjects, say physics, geography, biology or the humanities, will provide examples that require mathematical modelling so that students may develop a more open picture of mathematics. However, even if such examples were provided in other subjects, which does not seem to happen very often, except presumably in physics, this may not influence students' picture of mathematics because the examples would not necessarily be perceived as being mathematical in nature. Furthermore, some students enjoy dealing exclusively with concepts and theories inside mathematics, or they find joy or satisfaction in the successful performance of clearly defined algorithms, and regard aspects of the world outside mathematics as an inappropriate disturbance of their view of mathematics. It is a non-trivial and generally long-term task for the teacher to change these students' picture of mathematics towards a wider view that includes real-world connections. According to what is known from a multi-faceted set of experiences, the most promising way to achieve such a change is to treat a collection of extra-mathematical examples over a longer period and address the role of mathematics in these examples. Long-term studies such as Maaß (2004) have shown that it is indeed possible to change students' picture of mathematics in such a way.

5.3 Challenges on the teachers' side

What students need to be able to perform modelling activities holds even more for teachers. First, the teachers themselves must be able to carry out the modelling tasks they give to their students. This does not necessarily mean that they must have done so prior to task assignment, but teachers should be able to design or choose tasks which present challenges to their students that are suitable in terms of their backgrounds and prerequisites. Teachers must have at least the same mathematical and extra-mathematical knowledge needed to manage these tasks as expected from the students. Similarly, they must possess at least the same competencies and appropriate mathematics-related beliefs as the students. Second, modelling tasks make mathematics lessons more open, more demanding and less predictable not only for learners but also for teachers. Written and oral assessments involving modelling become more complex and more difficult to judge and grade. Teachers must learn how to deal with such situations and demands. When treating modelling cases, teachers must be alert to unexpected ideas or difficulties arising on the students' side and to be able to react flexibly to them. A lack of these competencies, necessary for teaching and assessing modelling, is perhaps the biggest barrier for the implementation of modelling in everyday classrooms or, as Freudenthal (1973, p. 73) expressed it:

> Among all the arguments against teaching mathematics that is not isolated from applications, I can understand that of incompetence, and if it is no affectation, I can appreciate it. The mathematics teacher does not know how mathematics is applied, and we cannot blame him for this ignorance. Where should he have learned it?

This "incompetence" (sometimes perhaps only a feeling of being incompetent) may refer to all components mentioned above: insufficient mathematical knowledge for dealing with certain modelling problems, insufficient extra-mathematical knowledge (particularly if the extra-mathematical modelling context does not refer to the mathematics teacher's other subject(s) of study), insufficient competencies for successfully dealing with modelling tasks, or insufficient pedagogical competencies.

Taking into account empirical results about the crucial role that teachers' competencies, in particular the subject-related ones, have for the quality of their instruction and the learning of their students (see Kunter et al., 2013; Schmidt et al., 2007), it is important to know as much as possible about which competencies are necessary for teaching modelling. In the previous paragraph, we listed some of them. The competencies listed in Borromeo Ferri and Blum (2010) comprise a "theoretical" dimension (including modelling cycles), a "task" dimension (including cognitive task analyses), an "instructional" dimension (including interventions and support), and a "diagnostic" dimension (including identifying students' difficulties); for more details, see section 6.6 and Borromeo Ferri (2017). All these elements must be discussed in teacher education and professional development since it is not realistic to expect that teachers will gain the necessary professional competencies and knowledge just from teaching practice. It is desirable to uncover, by means of empirical studies, what teacher competencies related to modelling, as well as other teacher competencies, are particularly influential with respect to the quality of instruction and eventually to students' learning and how strong the different effects are.

The most obvious consequence of these considerations is to include mathematical modelling and its teaching in the key phases of mathematics teacher education, also and especially in professional development (PD) activities for practising teachers. Effective in-service education is characterised by, among other aspects, the following criteria (cf. Zehetmeier & Krainer, 2011; Lipowsky & Rzejak, 2012; Roesken-Winter et al., 2015; Maaß & Engeln, 2018): organising a series of events with intermediate practical experiences (in contrast to "one afternoon" events); linking the PD content explicitly to learning and teaching in the classroom, both the teachers' own classrooms and reports or videos from other classrooms; stimulating participants' own activities in task construction, problem solving, lesson design, etc.; and stimulating reflections by confronting teachers' knowledge, beliefs and everyday classroom actions with those presented in the PD course. There are empirical indications that courses which pay attention to these criteria are likely to result in a change of teachers' beliefs about modelling and a subsequent change of students' perception of their teachers' teaching (see the report of Maaß & Engeln, 2018, about such a PD course in 12 European countries). It would be desirable to have well-defined and controlled studies on the effects of modelling courses for practising teachers on their professional knowledge, their actual teaching and their students' learning.

5.4 Challenges concerning classroom interaction and organisation

We mentioned in section 5.2 that modelling activities make lessons more open. More generally, they influence the didactical contract substantially. By "didactical contract" (*contrat didactique*, see Brousseau, 1997), we understand the set of habits, rules and expectations that are, explicitly or implicitly but mostly in a tacit manner, established between students and their teacher concerning their interaction and division of labour. The contract may well differ from teacher to teacher and from classroom to classroom, and it develops during a student's educational life. Among such contract rules and expectations one often finds the following (formulated from a student's perspective), many of which will be in conflict with the demands of genuine modelling activities:

1 The teacher only gives me tasks that can be completed within at most 10 minutes by means of the topic dealt with during the last few weeks. The task formulation includes questions that are clear and unambiguous, and tasks are always posed by the teacher.
2 The task contains exactly those data that are needed to answer the questions posed. It is never necessary to find additional information or data or to make assumptions beyond the ones stated in the presentation of the task. Similarly, none of the information or data provided is superfluous.
3 The tasks always have a unique answer; the teacher tells me if I am on a wrong track and gives me hints as to how to find this answer.
4 If the task has been set in a real-world context, it is typically unnecessary or even counter-productive to try to understand that context; instead, the contextual disguise has to be stripped off as quickly as possible to find the mathematical problem that the teacher wants me to solve.
5 The teacher controls my solution and tells me if something is wrong.

Of course, not only the intrinsic demands of modelling tasks but the whole pedagogy around dealing with them in teaching and learning requires a revision of such rules and expectations. Let us consider the "filling up" example (see example 2 in Chapter 3 where the question is whether it is worthwhile driving to a distant petrol station to fill up a car). In lessons involving this task, students must, first of all, take the context seriously and imagine and conceptualise the situation to be able to identify the relevant variables (in particular: distance, petrol consumption, tank volume) and relations between them and to interpret what "worthwhile" could or should mean. They do not know in advance what kind of mathematics will be needed to solve the task unless they have dealt with very similar tasks before. They will oftentimes not only work individually but also exchange ideas with their peers about their assumptions concerning the parameters of the car or about their modelling approaches. The students in the class are likely to produce several different models and answers that are compared, discussed and revised in the light of ideas from other students' solutions. The students are also expected to critically check their solutions and revise them if need be.

For a task such as "filling up", there will probably be sufficient time to treat the most basic solution ("worthwhile" means "cheaper") within a normal lesson so that such tasks can be integrated in the usual organisation of a school day. However, if further parameters (such as time, risk of accidents or air pollution) are to be considered, a normal or double lesson will not suffice. This is even more true of an example such as "traffic flow" (see example 6 in Chapter 3). Students need time to understand the problem, to identify the relevant variables among several possible options and to find a reasonable and useful model. In order to find realistic values for the parameters, they need data, so, in an "activity-oriented" learning environment, they ought to leave the classroom and observe actual traffic flows and record data about them, and they ought to validate their results by applying them in real traffic situations. Such modelling activities may easily conflict with the way schooling is typically organised, especially complex modelling problems. Among the problems of this kind are the following (see Kaiser et al., 2013): optimal capacity utilisation of airplanes, radio-therapy planning for cancer patients, identification of fingerprints, optimal position of rescue helicopters, pricing for internet booking of flights, optimisation of roundabouts, risk management, optimal arrangement of automatic water irrigation systems, prediction of the spread of a disease, optimal planning of a wind park, and planning of bus lines and bus stops. Such problems require students, among other things, to do their own investigations, to make simulations or even practical experiments, to construct competing models and to evaluate them in the real world.

Another source of challenges is mathematics curricula and syllabi. They typically require teachers to cover a certain number of topics within a limited span of time, so many teachers are afraid of "losing" too much time if and when treating modelling examples. Also, the pressure to have enough curricular subject matter available for the next class test may prevent teachers from attributing a substantial role to modelling tasks. The curricula may well state that the aim of mathematics teaching is not only, or not even primarily, to cover the set mathematical topics but also to develop students' conceptual understanding and mathematical competencies. However, it is not unusual for teachers to believe – perhaps unconsciously – that the competency to apply mathematics to extra-mathematical domains and to model real-world situations develops more or less automatically as a result of solely dealing with mathematical concepts, facts and procedures – which, as mentioned in the beginning of this chapter, unfortunately is not true (see section 5.1). Once again, the only way

98 Challenges for the implementation of mathematical modelling

to develop students' modelling competency is to involve them in modelling activities while taking into account significant criteria for quality teaching: an effective and learner-oriented classroom management; cognitive as well as meta-cognitive activation of learners; and a reflectively designed, challenging orchestration of the subject matter (for details see section 6.4). A productive syllabus leaves enough room for the development of mathematical competencies in close connection with mathematical knowledge and skills.

The inclusion of modelling and applications activities in mathematics teaching may take several forms. Blum and Niss (1991) distinguish between six types of approaches to such inclusion:

1. The "separation" approach where modelling activities take place in separate sections or courses and mathematics lessons are restricted to intra-mathematical activities;
2. The "two-compartment" approach where the whole course (say half a school year) is divided into two parts: the first part deals with pure mathematics and the second part uses the mathematics developed to treat models and modelling examples;
3. The "islands" approach where the whole course is divided into several segments, each organised according to the two-compartment approach;
4. The "mixing" approach where every time a new mathematical topic area is treated, extra-mathematical examples are used to assist the introduction and consolidation of the topics which are then, afterwards, applied to extra-mathematical situations;
5. The "curriculum integrated" approach where problems, also (but not only) from the real world, are employed to develop new mathematics (new to the students, of course), which is afterwards used to solve further problems, also from extra-mathematical domains; and
6. The "interdisciplinary integrated" approach where mathematics and extra-mathematical problems are integrated throughout the course.

In a usual school environment, genuinely integrated approaches are not often found. However, more important than the course organisation is the way the teaching and learning takes place. Within all the forms listed, successful teaching pays attention to the abovementioned well-founded criteria of quality teaching.

5.5 Challenges imposed by assessment

It is important to include modelling not only in instruction but also in achievement or diagnostic tests and in examinations (summative assessment) as well as in all kinds of feedback that teachers give to their students during teaching/learning processes (formative assessment). It is a general observation, also experienced by many teachers, that in an assessment-based system, only those demands which are assessed are taken seriously by the students, whether in classwork and/or in summative assessment, including final examinations – "What you assess is what you get!" (Burkhardt et al., 1990; Niss, 1993). It is certainly true that not everything that is taught can or should be assessed; on the other hand, it is not surprising that learners tend to reduce their learning load by concentrating mainly or exclusively on those topics and activities that actually are assessed.

The challenges raised by the very presence of summative or formative assessment with regard to applications and modelling are important enough, but the aspects that are not actually being assessed are equally important because they signal to students what they should emphasise, or let go unnoticed, in their modelling work.

This gives rise to the question of how modelling can be made an object of assessment, in particular tests and examinations (see Niss, 1993, for a conceptual framework). Since assessment oftentimes focuses on the individual, an assessment task must be formulated so students can understand it without support from others, can work on it individually and can complete it within a reasonable time frame. The last-mentioned criterion, in particular, can create substantial restrictions since genuine full-fledged modelling work usually requires quite some time, for instance, when information has to be sought, data have to be collected or experiments have to be conducted. That modelling can indeed be assessed has been shown in many studies during the last three decades. We mention here in particular the work of a British-Australian group, see Haines and Izard (1995), Haines et al. (2001), and Houston and Neill (2003), which has influenced the discussion on assessment of mathematical modelling since the 1990s (for more details, see the surveys by Frejd, 2013; Kaiser, 2017; also see section 6.8). The relevance of and the possible approaches to formative assessment of modelling have been emphasized by Eames et al. (2016). An example of formative assessment is the "Matchstick" problem (Figure 5.1) developed in the Shell Centre for Mathematical Education for the Mathematics Assessment Project (map.mathshell.com, see Swan & Burkhardt, 2014, for more details; see also section 7.6 for more examples from the Shell Centre).

The students are meant to work on this problem independently. The teacher is supplied with a list of possible challenges and difficulties to students. S/he collects students' work, makes an overall assessment of it (without scoring it) and gives qualitative feedback. In the following phase, the students discuss their work in small groups, prepare group solutions, compare their solutions with those of other groups and improve them once again. In the end, the group solutions are presented to the whole class, and the solution process is collectively reflected upon in retrospect.

An additional challenge is how students' completion of modelling tasks can be marked. This is different from marking intra-mathematical tasks where there are usually clear criteria of what counts as right or wrong, good or bad. A generic approach to marking students' responses to modelling tasks (see Leiß & Müller, 2008, for a concrete suggestion) is to identify the key elements of an ideal-typical modelling process providing answers to the task. We can, for instance, use the cognitive modelling cycle shown in section 2.5, sort a given student's modelling steps according to this cycle and assess every single step. Let us imagine that the task "Uwe Seeler's foot" (example 1 in Chapter 3) is given in a written test, where the

Making Matchsticks

Matchsticks are
1/10 inch by 1/10" by 2"

Matchsticks are often made from pine trees.

Estimate how many matchsticks can be made from this tree: 80 ft tall 2 ft diameter at the base.

You may find some of the information given on the formula sheet helpful.

Explain your work carefully, giving reasons for any choices you make.

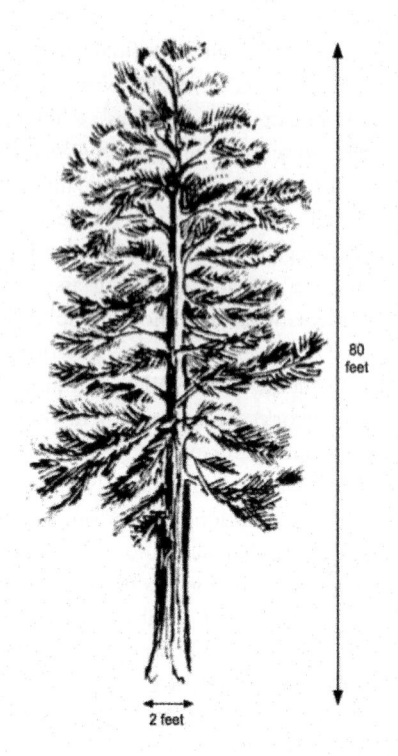

80 feet

2 feet

FIGURE 5.1 The Matchstick problem for formative assessment

students are asked to find the possible height of a statue of the German soccer player Uwe Seeler on the basis of the measures of a sculpture of his foot. When marking a student's written response, the teacher may try to identify the following elements:

1 "Understanding the situation": Are there indications that the issue of finding the height of a person based on his foot length has been understood?
2 "Simplifying and structuring the situation": Are the simplifications made by the student(s) appropriate; have the relevant variables and relations been identified; are the assumptions reasonable; is a sketch of the situation part of the response?
3 "Mathematising": Have the variables and relations been appropriately translated into an equation or a relation using proportions and scaling factors?
4 "Working mathematically": Are the mathematical calculations and considerations correct, and do they lead to an answer specifying a length?
5 "Interpreting": Has the mathematical result been correctly interpreted as the unknown height of the statue accompanied by a meaningful unit; is the value appropriately rounded; and is there a final answer?
6 "Validating and evaluating": Are there indications that the result has been validated and the model evaluated?
7 "Exposing": Are the solution process and its result coherently presented and clearly structured?

The first and the last step in the process, "Understanding" and "Validating and evaluating", are difficult to identify in a written report of the completed work since these steps often take place in the head of the modeller only. In oral assessments, these steps can be judged as well by inviting the respondent to explain how s/he approached the situation and how s/he checked the outcomes.

Some teachers might prefer to use points for single elements of the modelling work and add them at the end. However, more important than assigning and adding up points is identifying strengths and weaknesses of the work presented and giving appropriate feedback. Points might even distract students' attention from the feedback (see Black & Wiliam, 1998). Besser et al. (2013) suggest to give feedback in three parts: (1) strength ("You are already quite good at dealing with the following . . ."); (2) weaknesses ("You can still improve in dealing with the following, if you pay attention to my hints . . ."); and (3) hints ("Hints on how you can improve . . ."). This kind of feedback is corroborated by the results of empirical findings on effective feedback (see Hattie &Timperley, 2007).

It ought to be mentioned that providing high-quality assessment and feedback to students on their modelling work is demanding and time consuming. Here the idea of peer assessment may contribute to disburdening teachers (Black &Wiliam, 1998) and at the same time give students more responsibility for their work. In any case, it is a challenge for the education system to provide legislative and working conditions and resources for teachers to allow them to undertake such work. And it is a challenge for teachers to navigate within the boundary conditions they are under and find ways to provide quality assessment and feedback both to their students and to the system.

5.6 Challenges in finding suitable materials

One explanation of the absence of modelling in everyday classrooms often put forward is the lack of appropriate teaching materials. This explanation no longer seems valid considering the multitude of existing materials. In the last three decades, a plethora of valuable and feasible teaching materials (books, collections of examples, reports on experiences) have been published in many countries, including Australia, Denmark, Germany, Japan, the UK, and the USA. In the following section, we will refer to some of these materials.

Many examples of modelling tasks and learning environments for modelling can be found in the series of Proceedings of the ICTMA conferences. Under www.ictma.net/literature.html, all ICTMA books are listed. The organisation COMAP (the Consortium for Mathematics and its Applications, also see section 7.4) has produced, in the last four decades, materials containing modelling examples; see www.comap.com. We mention here the book *For All Practical Purposes* (the last edition of which is COMAP, 2013) and the GAIMME Report, produced in collaboration with SIAM (COMAP, 2016). The GAIMME Report contains a wide range of modelling examples from elementary to upper secondary level and from word problems to modelling projects; several examples are accompanied by detailed hints concerning how these examples can be used in the classroom. Since 1999, COMAP has organised two annual international modelling competitions for teams of college students, the Mathematical Contest in Modeling (MCM) and the Interdisciplinary Contest in Modeling (ICM). The modelling problems as well as possible solutions are available at www.comap.com/undergraduate/contest. Since 2015, COMAP, in cooperation with NeoUnion, has also organised an international modelling contest for teams of upper secondary school students, the International Mathematical Modeling Challenge (IM^{2}C). The modelling problems and possible solutions are available at www.immchallenge.org. The Australian branch of this contest has established its own website (www.immchallenge.org.au), which also contains the book by Galbraith and Holton (2018); also see section 7.3.

The Shell Centre for Mathematical Education at Nottingham University (also see section 7.4) has developed, over the last five decades, a multitude of modelling materials for the lower secondary level (for an overview, see Burkhardt, 2018). These materials comprise modules for modelling projects where students are expected to work in groups for several hours, developed by Malcolm Swan and others in the Numeracy Through Problem Solving project: "Design a board game", "Produce a quiz show", "Plan a trip", "Be a paper engineer", and "Be a shrewd chooser". In the Bowland Mathematics project (see www.bowlandmaths.org.uk), the Shell team developed modelling units expected to take four to five hours, such as "Reducing road accidents" and "How risky is life". In the Mathematics Assessment project, a collaboration between the Shell Centre and the University of California at Berkeley, the modelling lesson materials supported formative assessment for learning (see www.map.mathshell.org/lessons.php and also cf. section 5.5).

Another rich source of modelling examples is the collection of modelling activities developed by Lesh and his group within the framework of the "models and modelling perspective on teaching, learning, and problem solving" (see Lesh & Doerr, 2003, and section 6.9). Although primarily developed as a research tool, the model development sequences of this project can also be used for classrooms ranging from primary to upper secondary school levels, as well as for teacher education. A typical sequence consists of three parts: a "Model Eliciting Activity" (MEA, see section 2.7), followed by a "Model Exploration Activity" (MXA), and finally a "Model Application Activity" (MAA). Teaching methods for MEAs are described in detail in Moore et al. (2018); for more information on the design principles for these activities, see Brady et al. (2018).

Especially for the German speaking world, the series of books *Materialien für einen realitätsbezogenen Mathematikunterricht* (*Materials for Reality-Oriented Mathematics Teaching*), edited by the so-called German Istron group, presents a wealth of tried and tested modelling examples for all school grades, with an emphasis on the lower secondary level. Since 1993, 24 volumes have been published (see www.istron.mathematik.uni-wuerzburg.de/istron/index.html@p=1033.html for an overview). Particularly interesting is a recent volume (Siller et al., 2018), which is a "best of" collection from all previous Istron books (for details of the group and especially the "best of" volume, see section 7.4).

However, one issue is the availability of resource materials that teachers can find if they search for it. Another issue is the presence or absence of sections and tasks on modelling in prevalent textbooks. In this respect, the situation is very diverse within and across countries. Since it is well known from research that textbooks constitute the predominant teaching materials upon which teachers base their teaching, it is evident that textbooks without suitable modelling sections and tasks generate marked challenges to the implementation of modelling work in everyday teaching.

Another challenge is rooted in the fact that the modelling examples, tasks, units and materials found in different types of literature display a wide range of content and quality. If a teacher looks for modelling material suitable for a certain lesson or teaching unit, it may be difficult to find something that is appropriate with respect to the mathematical topics and the extra-mathematical contexts involved (if it is desirable at all to match curricular topics and modelling topics closely, see section 5.2). The contexts and the problems considered therein may range from purely dressed-up to authentic. Burkhardt (1981) classifies modelling problems as Action, Believable, Curious, Dubious and Educational. The notion of an "authentic problem" may have several different meanings (see Palm, 2002). The strictest meaning is that the context and the problem to be dealt with come directly from a genuine field of practice in industry, business, science, society or everyday life. There are certainly examples for all levels that satisfy this demand (see Kaiser et al., 2013). However, this demand is often too strict for educational purposes since authentic problems tend to go beyond the reach of school mathematics either in terms of the mathematics involved (be it differential or difference equations, functions of several variables or advanced probability distributions or stochastic processes, or discrete mathematics) or in terms of the knowledge needed from other fields (such as physics, biology, engineering or economics). A less strict notion of authenticity (see Niss, 1992; Palm, 2002, 2007; Vos, 2011, 2015) requires the contexts and problems to be constructed in such a way that they might occur in real practice and people from the practice area find them credible, albeit simplified. Crucial notions here are honesty and credibility (Carreira & Baioa, 2018), that is, the teacher ought to make it clear to the students in what respects a context or problem is not authentic in the strict sense but that it deserves to be taken seriously nevertheless. According to Vos (2018), criteria for authenticity in an educational context are: an out-of-school origin and a certification of originality (by suitable artefacts or by expert testimony). However, inauthentic contexts may in fact be suitable for educational purposes, depending on the teaching and learning goals (see section 2.7). For instance, when motivating the study or practise of certain mathematical topics is the primary aim, dressed-up problems might very well be appropriate, and the same holds when specific sub-competencies of modelling are to be practised or drilled. The question of how close a task context is to reality also has a subjective aspect. Students might hold a rather narrow view of authenticity, oriented towards their own personal life, and at the same time be more generous about the demand for authenticity since they know that instruction has multiple purposes and school is different from real life.

The more general question behind these considerations about the authenticity of tasks is what a "good" modelling task should look like. As has been said, this depends strongly on the goals the teacher wants to pursue by a given task. However, there are certain criteria that can be applied to judge whether a given task is suitable for the intended purposes. These criteria comprise, among others, the following (see Maaß, 2010; Borromeo Ferri, 2017):

1   Degree of precision or openness of the task question: Does the formulation of the question itself suggest a solution approach, or does a more precise question have to be developed during the solution process?
2   The kind of information given: Does the task contain more or less information than is necessary for dealing with it?
3   Complexity of the question: Is an approach immediately recognisable, and will one loop in the modelling cycle be sufficient to arrive at a satisfactory solution, or are different approaches, or several loops, likely to be necessary?
4   The kind of real-world context: Is the context accessible and understandable by the students dealing with the task? Is the context credible, and is it relevant for students' lives?
5   Extent and level of the mathematical content: What kind of mathematics is suitable or necessary for solving the task? Is this mathematics accessible to the students?

5.7   Challenges concerning digital tools

The existence of digital tools (such as calculators, computers, tablets or smartphones) with powerful software can be a vehicle for enhancing and improving the treatment of modelling tasks. As an example, let us consider the modelling task "traffic flow" (see example 6 in Chapter 3). As soon as the situation is understood and the relevant variables – speed v, car length l and distance between cars, d – have been identified, computations of the flow rate F can be carried out with multiple varying speeds, car lengths and distance rules before the general model $F = v/(l + d)$ is developed. The real-world knowledge necessary for assigning concrete values or rules to the variables (car lengths, distance) can be extracted from the internet. It might also be helpful to simulate the real situation dynamically using various values of the variables. Once the general model has been established, a digital tool can be used to explore, both numerically and graphically, different functions $F_i = f_i(v)$, depending on the chosen distance rules, for various car lengths. The maxima of F_i for quadratic distance rules can be determined with the help of a Computer Algebra System (CAS) tool, both numerically and symbolically. When comparing and evaluating different models, calculations and visualisations will help form an image of what is going on.

This example suggests that digital tools can indeed be used as powerful aids for modelling activities, in all phases of the modelling process, not only in the intra-mathematical treatment phase (see, e.g., Drijvers, 2003; Borba & Villarreal, 2005;

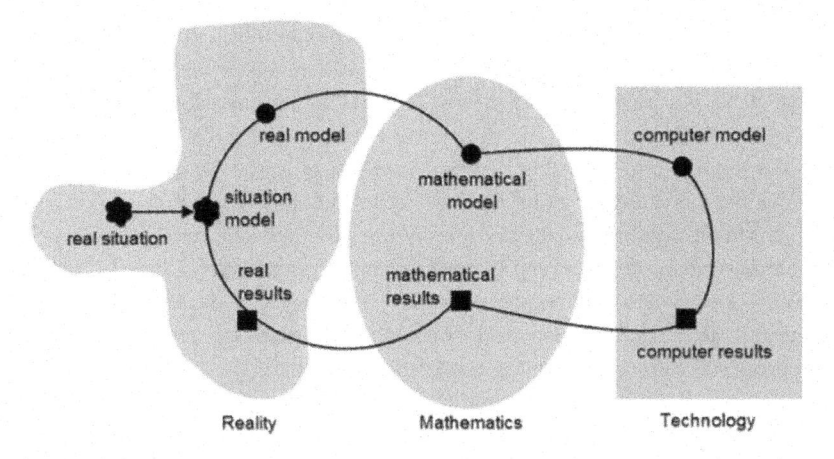

[figure: extended modelling cycle diagram with labels — real situation, situation model, real model, real results (Reality); mathematical model, mathematical results (Mathematics); computer model, computer results (Technology)]

FIGURE 5.2 The extended modelling cycle
Source: Greefrath, 2011, p. 302

Henn, 2007; Confrey & Malony, 2007; Geiger, 2011; Greefrath et al., 2011; Daher & Shahbari, 2015; Greefrath & Siller, 2017; Greefrath et al., 2018). Relevant activities include experiments, investigations, simulations, visualisations and calculations. In order to visualise the use of digital tools in the intra-mathematical phase of the modelling cycle, Greefrath (2011) suggests extending the cycle by adding a third world beside the extra-mathematical and the mathematical worlds: the technological world (Figure 5.2).

A special feature of a many digital tools is the availability of statistical packages, which are of particular relevance to mathematical modelling. These tools typically allow for parameter estimation, hypothesis testing and – especially – regression analysis, which are significant aspects of modelling involving real data, aspects which usually are very time consuming if they have to be dealt with manually by the modeller. In addition, ordinary spreadsheets can handle most kinds of data very effectively.

There are many empirical studies which show that digital tools can actually enhance the acquisition of competencies in modelling environments; see Greefrath et al. (2018) both for an overview and for a quantitative study where the experimental group used a dynamic geometry software (DGS). In this study, the control group, which only used paper and pencil, had modelling competency gains comparable to those of the experimental group. One reason might be that the test tasks in this study could be solved without DGS, whereas an advantage of the use of DGS will only be visible with tasks where DGS allows for new approaches. Other studies where DGS was successfully used to support modelling processes include Carreira et al. (2013) and Gallegos and Rivera (2015).

The existence of digital tools produces not only new opportunities but also new challenges to the implementation of modelling in mathematics classrooms. Students might, for instance, tend to use such a tool in the "traffic flow" example rather extensively for a multitude of calculations (flow rate for several speeds and a popular distance rule) and might get lost in data instead of carefully analysing the situation without digital support (What happens if the distance increases linearly with the speed? What if it increases quadratically? What is the flow rate for very small speeds, and what is it for very large speeds in these two cases?). How will the understanding of a functional relationship between variables, as in the "traffic flow" example, be influenced by an early global visualisation of the underlying function on a huge domain? Such a global visualisation presupposes the idea of a function as an object, which may not be appropriated by lower secondary students.

What knowledge and competencies are necessary if a digital tool can do all numerical, algebraic and analytical calculations; all graphical representations; and, if appropriate, simulations much more effectively than human problem solvers? To interpret the digital output appropriately, understanding and familiarity with those calculations, visualisations and simulations are necessary; otherwise, the tools might lead to unsuitable results which the user may uncritically accept. Some classical manipulative proficiencies are certainly becoming less important. On the other hand, the basic skills associated with a given tool will have to be acquired before the tool can be used appropriately, as is the case with any device: The device itself produces new demands for learning. The availability of tools which can easily do numerical or algebraic calculations and which can supply any kind of functions for modelling given data might lead to an inappropriate emphasis on quantitative approaches, neglecting qualitative relationships and especially reflections on which approaches make sense in the context of the given problem situation. Teachers must carefully consider the advantages and disadvantages of the use of digital tools in modelling work and determine when and how they should – or should not – use such tools.

References

Besser, M., Blum, W. & Klimczak, M. (2013). Formative assessment in everyday teaching of mathematical modelling: Implementation of written and oral feedback to competency-oriented tasks. In: G. Stillman, G. Kaiser, W. Blum & J. Brown (Eds.), *Teaching Mathematical Modelling: Connecting to Research and Practice* (pp. 469–478). Dordrecht: Springer.

Black, P. & Wiliam, D. (1998). Assessment and classroom learning. In: *Assessment in Education: Principles, Policy & Practice* **5**(1), 7–74.

Blum, W. (2015). Quality teaching of mathematical modelling: What do we know, what can we do? In: S. J. Cho (Ed.), *The Proceedings of the 12th International Congress on Mathematical Education: Intellectual and Attitudinal Challenges* (pp. 73–96). New York: Springer.

Blum, W. & Niss, M. (1991). Applied mathematical problem solving, modelling, applications, and links to other subjects: State, trends and issues in mathematics instruction. In: *Educational Studies in Mathematics* **22**(1), 37–68.

Borba, M. & Villarreal, M. E. (2005). *Humans-with-Media and the Reorganization of Mathematical Thinking: Informations and Communication Technologies, Modeling, Experimentation and Visualization.* New York: Springer.

Borromeo Ferri, R. (2017). *Learning How to Teach Mathematical Modeling in School and Teacher Education.* Cham: Springer.

Borromeo Ferri, R. & Blum, W. (2010). Mathematical modelling in teacher education: Experiences from a modelling seminar. In: V. Durand-Guerrier, S. Soury-Lavergne & F. Arzarello (Eds.), *CERME-6: Proceedings of the Sixth Congress of the European Society for Research in Mathematics Education* (pp. 2046–2055). Lyon: INRP.

Brady, C., Eames, C. & Lesh, R. (2018). The student experience of model development activities: Going beyond correctness to meet a client's needs. In: S. Schukajlow & W. Blum (Eds.), *Evaluierte Lernumgebungen zum Modellieren* (pp. 73–92). Wiesbaden: Springer Spektrum.

Brousseau, G. (1997). *Theory of didactical situations in mathematics*. Dordrecht: Kluwer.

Burkhardt, H. (1981). *The real world and mathematics*. Glasgow: Blackie.

Burkhardt, H. (2004). Establishing modelling in the curriculum: barriers and levers. In: H.-W. Henn & W. Blum (Eds.), *ICMI Study 14: Applications and Modelling in Mathematics Education Pre-Conference Volume* (pp. 53–58). Dortmund: University of Dortmund.

Burkhardt, H. (2018). Ways to teach modelling: A 50 year study. In: *ZDM: The International Journal on Mathematics Education* **50**(1 + 2), 61–75.

Burkhardt, H., Fraser, R. & Ridgway, J. (1990). The dynamics of curriculum change. In: I. Wirszup & R. Streit (Eds.), *Developments in School Mathematics Around the World 2* (pp. 3–30). Reston: National Council of Teachers of Mathematics.

Carreira, S., Amado, N. & Canário, F. (2013). Students' modelling of linear functions: How GeoGebra stimulates a geometrical approach. In: B. Ubuz et al. (Eds.), *CERME 8: Proceedings of the Eighth Congress of the European Society of Research in Mathematics Education* (pp. 1031–1040). Ankara: Middle East Technical University.

Carreira, S. & Baioa, A.M. (2018). Mathematical modelling with hands-on experimental tasks: On the student's sense of credibility. In: *ZDM: The International Journal on Mathematics Education* **50**(1/2), 201–215.

COMAP (Ed.) (2013). *For All Practical Purposes: Mathematical Literacy in Today's World* (10th Edition). New York: Freeman.

COMAP (Ed.) (2016). *Guidelines for Assessment & Instruction in Mathematical Modelling Environments*. Bedford: COMAP and Philadelphia: SIAM.

Confrey, J. & Malony, A. (2007). A theory of mathematical modelling in technological settings. In: W. Blum, P.L. Galbraith, H.-W. Henn & M. Niss (Eds.), *Modelling and Applications in Mathematics Education: The 14th ICMI Study* (pp. 57–68). New York: Springer.

Daher, W. & Shahbari, A. (2015). Pre-service teachers' modelling processes through engagement with model eliciting activities with a technological tool. In: *International Journal of Science and Mathematics Education* **13**(Suppl 1), 25–46.

De Lange, J. (1987). *Mathematics, Insight and Meaning*. Utrecht: CD-Press.

Drijvers, P. (2003). Algebra on screen, on paper, and in the mind. In: J.T. Fey, A. Cuoco, C. Kieran, L. McMullin & R.M. Zbiek (Eds.), *Computer Algebra Systems in Secondary School Mathematics Education* (pp. 241–268). Reston: NCTM.

Eames, C., Brady, C. & Lesh, R. (2016). Formative self-assessment: A critical component of mathematical modeling. In: C. Hirsch & A. Roth McDuffie (Eds.), *Mathematical Modeling and Modeling Mathematics* (pp. 229–237). Reston: National Council of Teachers of Mathematics.

Frejd, P. (2013). Modes of modelling assessment – A literature review. In: *Educational Studies in Mathematics* **84**(3), 413–438.

Freudenthal, H. (1973). *Mathematics as an Educational Task*. Dordrecht: Reidel.

Galbraith, P. & Holton, D. (2018). *IM^{2}C: Mathematical Modelling: A Guide Book for Teachers and Teams*. Melbourne: ACER.

Gallegos, R.R. & Rivera, S.Q. (2015). Developing modelling competencies through the use of technology. In: G.A. Stillman, W. Blum & M.S. Biembengut (Eds.), *Mathematical Modelling in Education Research and Practice* (pp. 443–452). Cham: Springer.

Geiger, V. (2011). Factors affecting teachers' adoption of innovative practices with technology and mathematical modelling. In: G. Kaiser, W. Blum, R. Borromeo Ferri & G. Stillman (Eds.), *Trends in Teaching and Learning of Mathematical Modelling (ICTMA 14)* (pp. 305–314). Dordrecht: Springer.

Greefrath, G. (2011). Using technologies: New possibilities of teaching and learning modelling: Overview. In: G. Kaiser, W. Blum, R. Borromeo Ferri & G. Stillman (Eds.), *Trends in Teaching and Learning of Mathematical Modelling, ICTMA 14* (pp. 301–304). Dordrecht: Springer.

Greefrath, G., Hertleif, C. & Siller, H.-S. (2018). Mathematical modelling with digital tools: A quantitative study on mathematising with dynamic geometry software. In: *ZDM: The International Journal on Mathematics Education* **50**(1/2), 233–244.

Greefrath, G. & Siller, H.-S. (2017). Modelling and simulation with the help of digital tools. In: G.A. Stillman, W. Blum & G. Kaiser (Eds.), *Mathematical Modelling and Applications, ICTMA 17* (pp. 529–539). Dordrecht: Springer.

Greefrath, G., Siller, H.-S. & Weitendorf, J. (2011). Modelling considering the influence of technology. In: G. Kaiser, W. Blum, R. Borromeo Ferri & G. Stillman (Eds.), *Trends in Teaching and Learning of Mathematical Modelling (ICTMA 14)* (pp. 315–329). Dordrecht: Springer.

Haines, C.R., Crouch, R. & Davis, J. (2001). Understanding students' modelling skills. In: J. Matos, W. Blum, K. Houston & S. Carreira (Eds.), *Modelling and Mathematics Education, ICTMA 9: Applications in Science and Technology* (pp. 366–380). Chichester: Horwood.

Haines, C.R. & Izard, J. (1995). Assessment in context for mathematical modelling. In: C. Sloyer, W. Blum & I. Huntley (Eds.), *Advances and Perspectives in the Teaching of Mathematical Modelling and Applications* (pp. 131–149). Yorklyn: Water Street Mathematics.

Hattie, J. & Timperley, H. (2007). The power of feedback. In: *Review of Educational Research*, **77**(1), 81–112.

Henn, W. (2007). Modelling pedagogy: Overview. In: W. Blum, P.L. Galbraith, H.-W. Henn & M. Niss (Eds.), *Modelling and Applications in Mathematics Education: The 14th ICMI Study* (pp. 321–324). New York: Springer.

Houston, K. & Neill, N. (2003). Assessing modelling skills. In: S.J. Lamon, W.A. Parker & S.K. Houston (Eds.), *Mathematical Modelling: A Way of Life: ICTMA 11* (pp. 155–164). Chichester: Horwood.

Ikeda, T. (2007). Possibilities for, and obstacles to teaching applications and modelling in the lower secondary levels. In: W. Blum, P.L. Galbraith, H.-W. Henn & M. Niss (Eds.), *Modelling and Applications in Mathematics Education* (pp. 457–462). New York: Springer.

Kaiser, G. (2017). The teaching and learning of mathematical modeling. In: J. Cai (Ed.), *Compendium for Research in Mathematics Education* (pp. 267–291). Reston: NCTM.

Kaiser, G., Bracke, M., Göttlich, S. & Kaland, C. (2013). Authentic complex modelling problems in mathematics education. In: A. Damlamian, J.F. Rodrigues & R. Sträßer (Eds.), *Educational Interfaces between Mathematics and Industry* (pp. 287–297). New York: Springer.

Kunter, M., Baumert, J., Blum, W., Klusmann, U., Krauss, S. & Neubrand, M. (Eds.) (2013). *Cognitive Activation in the Mathematics Classroom and Professional Competence of Teachers: Results from the COACTIV Project*. New York: Springer.

Leiß, D. & Müller, M. (2008). Offene Aufgaben – auch ein offenes Problem der Bewertung? In: *Praxis Schule 5–10* **19**(5), 13–17.

Lesh, R.A. & Doerr, H.M. (2003). *Beyond Constructivism: A Models and Modelling Perspective on Teaching, Learning, and Problem Solving in Mathematics Education*. Mahwah: Lawrence Erlbaum.

Lipowsky, F. & Rzejak, D. (2012). Lehrerinnen und Lehrer als Lerner – Wann gelingt der Rollentausch? Merkmale und Wirkungen wirksamer Lehrerfortbildungen. In: *Schulpädagogik heute* **3**(5), 1–17.

Maaß, K. (2004). *Mathematisches Modellieren im Unterricht – Ergebnisse einer empirischen Studie*. Hildesheim: Franzbecker.

Maaß, K. (2010). Classification scheme for modelling tasks. In: *Journal für Mathematik-Didaktik* **31**, 285–311.

Maaß, K. & Engeln, K. (2018). Impact of professional development involving modelling on teachers and their teaching. In: *ZDM: The International Journal on Mathematics Education* **50**(1 + 2), 273–285.

Moore, T., Doerr, H. & Glancy, A. (2018). Approaches to teaching mathematics through modeling. In: R. Borromeo Ferri & W. Blum (Eds.), *Lehrerkompetenzen zum Unterrichten mathematischer Modellierung* (pp. 215–231). Wiesbaden: Springer.

Niss, M. (1992). Applications and modelling in school mathematics: Directions for future development. In: I. Wirszup & R. Streit (Eds.), *Proceedings of the Third UCSMP International Conference on Mathematics Education, October 30–November 1, 1991* (Vol. 3, pp. 346–361). Chicago: The University of Chicago School Mathematics Project.

Niss, M. (Ed.) (1993). *Investigations into Assessment in Mathematics Education*. Dordrecht: Kluwer.

Niss, M. (1994). Mathematics in society. In: R. Biehler, R. Scholz, R. Sträßer & B. Winkelmann (Eds.), *The Didactics of Mathematics as a Scientific Discipline* (pp. 367–378). Dordrecht: Kluwer.

Palm, T. (2002). *The Realism of Mathematical School Tasks: Features and Consequences*. Umeå: Umeå University.

Palm, T. (2007). Features and impact of the authenticity of applied mathematical school tasks. In: W. Blum, P.L. Galbraith, H.-W. Henn & M. Niss (Eds.), *Modelling and Applications in Mathematics Education: The 14th ICMI Study* (pp. 201–208). New York: Springer.

Pollak, H. (1979). The interaction between mathematics and other school subjects. In: UNESCO (Ed.), *New Trends in Mathematics Teaching IV* (pp. 232–248). Paris: UNESCO.

Roesken-Winter, B., Hoyles, C. & Blömeke, S. (2015). Evidence-based CPD: Scaling up sustainable interventions. In: *ZDM: The International Journal on Mathematics Education* **47**(1), 1–12.

Schmidt, B. (2009). Modelling in the classroom: Motives and obstacles from the teacher's perspective. In: V. Durand-Guerrier, S. Soury-Lavergne & F. Arzarello (Eds.), *CERME-6: Proceedings of the Sixth Congress of the European Society for Research in Mathematics Education* (pp. 2066–2075). Lyon: INRP.

Schmidt, W.H., Tatto, M.T., Bankov, K., Blömeke, S., Cedillo, T., Cogan, L. et al. (2007). *The Preparation Gap: Teacher Education for Middle School Mathematics in Six Countries (MT21 Report)*. East Lansing: MSU Center for Research in Mathematics and Science Education.

Siller, H.-S., Greefrath, G. & Blum, W. (Eds.) (2018). *Neue Materialien für einen realitätsbezogenen Mathematikunterricht 4. 25 Jahre ISTRON-Gruppe – eine Best-of-Auswahl aus der ISTRON-Schriftenreihe*. Wiesbaden: Springer Spektrum.

Swan, M. & Burkhardt, H. (2014). Lesson design for formative assessment. In: *Educational Designer* **2**(7). www.educationaldesigner.org/ed/volume2/issue7/article24/

Treilibs, V., Burkhardt, H. & Low, B. (1980). *Formulation Processes in Mathematical Modelling*. Nottingham: Shell Centre Publications.

Vos, P. (2011). What is "authentic" in the teaching and learning of mathematical modelling? In: G. Kaiser, W. Blum, R. Borromeo Ferri & G. Stillman (Eds.), *Trends in Teaching and Learning of Mathematical Modelling, ICTMA 14* (pp. 713–722). Dordrecht: Springer.

Vos, P. (2015). Authenticity in extra-curricular mathematics activities: Researching authenticity as a social construct. In: G. Stillman, W. Blum & M.S. Biembengut (Eds.), *Mathematical Modelling in Education Research and Practice: Cultural, Social and Cognitive Influences* (pp. 105–114). New York: Springer.

Vos, P. (2018). "How real people really need mathematics in the real world": Authenticity in mathematics education. In: *Education Sciences* **8**, 195–208.

Zehetmeier, S. & Krainer, K. (2011). Ways of promoting the sustainability of mathematics teachers' professional development. In: *ZDM: The International Journal on Mathematics Education* **43**(6–7), 875–887.

6

WHAT WE KNOW FROM EMPIRICAL RESEARCH – SELECTED FINDINGS ON THE TEACHING AND LEARNING OF MATHEMATICAL MODELLING

In this chapter, we will summarise important findings on the learning and teaching of mathematical modelling, with an emphasis on empirical research. However, since empirical research does not exist in a vacuum but depends on conceptual, analytical and theoretical ideas, notions and approaches, some of the material presented in this chapter goes beyond empirical findings to also encompass, for example, *a priori* considerations and hypothetical explanations of empirical observations.

In view of the existing body of literature on such findings, it is impossible to be exhaustive in reviewing the empirical research in the field. Therefore, this chapter represents a selection of findings. The selection has been guided by our wish to shed light on the barriers to the learning of modelling and on ways to overcome these barriers. Most of the empirical studies mentioned in this chapter are either qualitative small-scale studies or medium-sized (up to a few hundred subjects) qualitative-quantitative studies. However, reference will also be made to quantitative large-scale studies, although there are not many that have modelling as their primary focus. Since empirical research on the teaching and learning of mathematics is unequally distributed across countries, research from some countries will figure more prominently than that of others (thus, in the literature review by Stohlmann et al., 2016, most research contributions came from Germany). Moreover, the backgrounds of the authors have, of course, influenced the selection of findings reported here. Other authors have previously attempted to survey the current state of empirical research in mathematical modelling. Examples include Stillman (2015, 2019), Kaiser (2017), and Schukajlow, Kaiser and Stillman (2018), which in addition to being a survey paper in itself is also an introduction to a special ZDM issue (no. 50, 2018) on this topic.

Knowledge about such findings will be helpful for teachers so they can best help students develop the desired knowledge and competencies. All findings have been obtained in certain organisational, societal and cultural environments, and in each case it must be considered to what extent these findings can be generalised to other environments (see Burkhardt & Schoenfeld, 2003, for such methodological questions). This is a particular instance of a phenomenon which is certainly among the most important general issues and aspects of human learning, which we will discuss in the first section of this chapter: situatedness of cognition and transfer of learning.

6.1 Situated cognition and the issue of transfer

The very reason that we have general school education in all countries is that we believe in and have evidence of the possibility of learning certain things in one context and situation and then transferring the learning outcomes to make sense and be relevant in other contexts and situations. Were it only possible to employ what has been learned in circumstances that are exactly the same as the original ones, we would only have context- and situation-specific atomistic drills and rehearsed practice, not education in a more general sense. There is certainly nothing wrong with very focused and specific drills and rehearsed practice. On the contrary, this is essential in many vocations, occupations and professions. For a chef to learn to make a *soufflé au Grand Marnier* to perfection, for a mason to build a brick wall accurately according to specifications, for a pilot to competently fly an aeroplane of type X, and for an orthopaedic surgeon to successfully replace a hip in every patient he or she receives for such an operation, a huge number of hours of strongly focused and careful drills and practice are needed for satisfactory results to be guaranteed every time. Only then can the chef, the mason, the pilot or the surgeon consider gradually transferring the skills they have obtained to new situations (to, say, a *soufflé au chocolat*, a free standing brick arch, an aeroplane of type Y, or a knee) that deviate more or less from the original one. But even in such rigidly defined contexts, transfer to close but unrehearsed territory will, most probably, not be automatic and will give rise to challenges.

So, the key issues for education are: To what extent is learning situated and dependent on the context and situation in which it first took place? What are the conditions and characteristics of such situatedness, and how does this depend on the nature and substance of what has been learnt and on the learner? To what extent is the transfer of what has been learnt under one set of circumstances to be applicable in new sets of circumstances possible? What are necessary or sufficient conditions for such a transfer to take place? And how can education systems, institutions, researchers and teachers, among others, foster and further transfer?

Although the authors cannot provide an in-depth account of all these aspects and issues, research evidence suggests that learning in general, and cognition in particular, is more situated (see, e.g., Brown et al., 1989) and that transfer is much less automatic and more difficult to obtain than one might originally think.

For "situated cognition" and "transfer" specifically in *mathematics* (see Watson, 1998), every topic of learning carries with it "indices" referring to its specific learning context, where "context" is to be understood in a broad sense, including the specific learning environment, the specific mathematical topic and the specific intra- or extra-mathematical problem context. Here, too, independent and automatic transfer from one context to another cannot be expected. Human brains seem to store learning results in specific compartments without automatically linking these with other compartments and without automatically recognising existing structural similarities between different compartments. This is particularly relevant for any learning in the field of relations between the extra-mathematical world and mathematics. For empirical results about "modelling competency" (see Chapter 4 for this construct), we must equip this construct with several indices, especially referring to the mathematical topics and the extra-mathematical contexts involved. One may ask: What can "modelling competency" possibly mean if it is dependent on the learning context? Should we speak of several specific "modelling competencies", indexed by specific mathematical topics and extra-mathematical contexts?

Three issues are essential here. The first is: To what extent are competencies and knowledge, gained solely by working in intra-mathematical contexts, sufficient for successfully undertaking mathematical modelling? Or differently put, to what extent can intra-mathematical competencies and knowledge automatically be transferred to becoming modelling competencies and knowledge? The second is: To what extent is it possible to transfer modelling competencies and knowledge gained in one modelling context and situation to other modelling contexts and situations, and how does this depend on the characteristics of the contexts and situations at issue, including the "distance" from the initial to the new contexts and situations? The third issue is: To what extent is it possible to design teaching and learning activities that efficiently support transfer between modelling contexts and situations? Clearly, it is not possible to provide exhaustive and detailed answers concerning these issues within a single book chapter. In what follows, some of the answers are of a sketchier nature than others.

Concerning the first of these questions, it is sometimes said that it is not only necessary but also sufficient to have a broad knowledge and a deep insight concerning pure mathematics to be able to apply this knowledge in extra-mathematical situations, in other words to perform modelling activities. Consequently, it should not be necessary to devote special time and energy to modelling activities in the mathematics classroom. However, many empirical investigations have shown (see the surveys by DeCorte et al., 1996; Niss, 1999) that it is very difficult to transfer knowledge and skills from one context or task to a different context or task and that such transfer actually does not happen automatically but has to be deliberately catered to and organised. This implies that it cannot be expected that intra-mathematical knowledge acquired by someone can be successfully employed in extra-mathematical problem contexts without focused practising. In fact, several studies (e.g., Plath & Leiß, 2018) have found only a weak or no correlation between school marks in mathematics and modelling achievement; in many places, school marks are mostly given according to intra-mathematical skills and abilities. The results of OECD's so-called PISA studies (PISA stands for Programme for International Student Assessment and every three years tests the ability of 15-year-olds to solve contextual tasks by means of school mathematics; for the latest PISA results, see OECD, 2016, 2019) supply many examples of the difficulty of transfer from pure

REVOLVING DOOR

A revolving door includes three wings which rotate within a circular-shaped space. The inside diameter of this space is 2 metres (200 centimetres). The three door wings divide the space into three equal sectors. The plan below shows the door wings in three different positions viewed from the top.

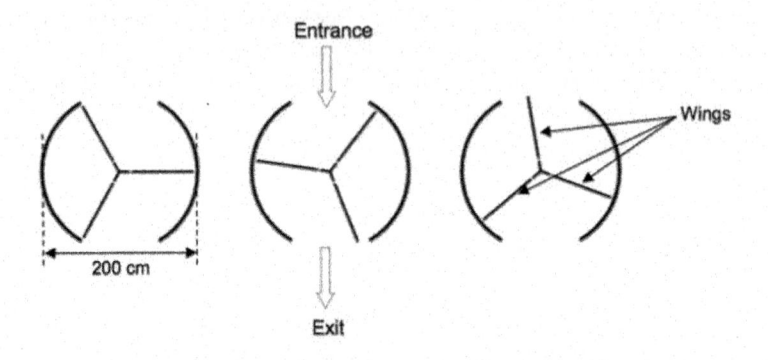

Entrance

⇩

200 cm   Wings

⇩

Exit

Question 1: REVOLVING DOOR   PM995Q01 – 0 1 9

What is the size in degrees of the angle formed by two door wings?

Size of the angle: ....°

FIGURE 6.1 PISA Item "Revolving door 1"

mathematical knowledge and skills to extra-mathematical contexts for all participating countries. As an example, in the task "Revolving door 1" (Figure 6.1; see OECD, 2013), students' knowledge about angles must be applied to a specific real-world situation described by a text (which also contains some information not needed for this first question) and three diagrams. Most participating students (15-year-olds) will know that a full circle is an angle of 360°, and they will also be able to divide 360 by 3. However, only 57.7% of the students in OECD countries could solve this task, that is, apply their well-established knowledge to this situation.

The difficulty of transfer (cf. the second issue) also appears when mathematical knowledge and modelling work that has been applied in a certain problem context is to be applied in a different context (Stillman, 1998, 2000). Even a relatively close transfer between structurally similar problems cannot be expected from students without support. For instance, students in a ninth class in Germany (in the framework of the DISUM project; see Blum & Leiß, 2007) had dealt with the "filling up" task (see example 2 in Chapter 3) and, supported by the teacher, successfully developed the basic solution. In a subsequent test, the students had to solve two structurally similar tasks: deciding whether it is worthwhile to drive to a nearby strawberry field to pick the berries for a cake instead of buying them in a nearby supermarket and whether it is worthwhile to use cloth diapers instead of disposable ones, which, in addition to the purchase costs, also generate washing costs. For many students, these tasks were totally new challenges, now about strawberries and diapers instead of cars, and consequently they had the very same problems as when they first began dealing with the "filling up" task. Such a structural similarity between tasks only becomes visible once the underlying common real or mathematical model has been constructed and students are made conscious of this commonality. We can expect more experienced modellers to see such similarities immediately because they are able to construct the underlying common model right away.

What we also know from research on situated cognition (see, e.g., Anderson et al., 1996) is that there is a chance for transfer if this transfer is explicitly organised and made conscious to students on a meta-level. For the "filling up" example and its related tasks, this means that to enable students to solve structurally similar tasks like the two test tasks mentioned above, this structural similarity has to be made explicit for the students by means of examples given by the teacher or even constructed by the students themselves. For instance, it can be shown why the problem of deciding whether it is worthwhile to drive to a distant beverage store in order to buy mineral water because it is cheaper there than in a nearby shop is "the same" problem as that of deciding whether it is worth driving to Luxembourg to fill up a car or whether it is worth using cloth diapers instead of disposable diapers; teachers can also demonstrate that the common mathematical modelling core of all these problems is to compare the values of a proportional and a linear function, even though the parameters and the variables involved stand for completely different things in the three different contexts. This could be done, for example, by writing down the solutions of the various problems in parallel and showing explicitly which variables, parameters and relations correspond to one another. This still does not guarantee transfer, which also depends on individual prerequisites and experiences, but at least it becomes more likely to occur if such measures are taken.

6.2 Barriers in the process of dealing with modelling tasks

Significant barriers are found in the processes involved in dealing with modelling tasks (compare the examples in Chapter 3). An analysis of these processes reveals that there are many possibilities for students to get stuck or to make mistakes when dealing with such tasks in the classroom or on examinations and tests. The first hurdle may already occur when students are invited to engage in a modelling task. This hurdle was found, for instance, in a study of prospective teachers who – in their capacity as learners – were reluctant to undertake mathematical modelling to deal with real-life problems concerning travelling in Greece (Potari, 1993). Jankvist and Niss (2019) found in a study of 315 upper secondary school students in Denmark that several students decline engagement in modelling tasks – either because they don't accept the task as being part of mathematics or because they have no idea of how to get started.

Mathematical modelling is a cognitively demanding activity (see Chapters 2 and 4) since several competencies must be activated along the road and both mathematical and extra-mathematical knowledge is required, as are appropriate beliefs and attitudes. Every step of the modelling process is a potential source of difficulty for students, and empirical research shows that every step can indeed be a barrier or blockage for some students (e.g., Ikeda & Stephens, 1998; Galbraith & Stillman, 2006; Stillman et al., 2010; Ludwig & Xu, 2010; Stillman, 2011; Blum, 2011; Schaap et al., 2011; Schukajlow et al., 2012).

In the following, we orient ourselves towards the "cognitive modelling cycle" presented in Figure 2.8, which describes, in an ideal-typical way, the steps that are involved when carrying out modelling tasks, and we will discuss every step separately. We presuppose that there is a task given to students consisting of an extra-mathematical situation together with certain questions. This task may have been set by the teacher or, after exploring some context, by the students themselves.

The *first step* that students must undertake when dealing with the task is "understanding the given situation and constructing a mental model of it". One factor influencing the way in which, and also how successfully, this can be done is the real-world context in which the situation is embedded (for different meanings of the word "context" in modelling, see Brown, 2019). Busse (2011) showed in a qualitative empirical study with 16- and 17-year-olds how idiosyncratically these students interpret the real-world context, depending on their previous personal experiences, and that familiarity with a context may sometimes even be obstructive because it may lead to representations and interpretations that are not helpful (see also Boaler, 1993). Several empirical studies have shown a big influence of an appropriate situation model (see Figure 2.8) on the solution rate of modelling tasks (see, e.g., Leiss et al., 2010). An important factor influencing the comprehension of a given task is students' language proficiency (as shown, e.g., by Plath & Leiß, 2018, in a study with seventh and eighth graders). Hence, this first step can be a big hurdle, and many students get stuck here. This fact does not only, or perhaps not even primarily, have cognitive roots. Many students around the world have learned, as part of the hidden school mathematics curriculum, that they may easily survive in mathematics without the effort of careful reading and understanding given mathematical tasks cast in an extra-mathematical context. It is part of their picture of mathematics that in this subject you don't have to care about meaning but just about tackling tasks by applying a recently learnt method and that a typical task contains exactly the information that is needed for finding the uniquely determined answer to it, neither more nor less. In the case of contextual problems, such students follow a "substitute strategy": "Ignore the context, just extract all data from the text and calculate something according to a familiar schema" (see, e.g., Nesher, 1980; Baruk, 1985; Schoenfeld, 1991; Lave, 1992; Reusser & Stebler, 1997; Verschaffel et al., 2000; Xin et al., 2007; Verschaffel et al., 2010). Schoenfeld and Verschaffel speak of the "suspension of sense-making" when playing the "word problem game".

Figure 6.2 shows an example (from the DISUM project), a student's answer to the "filling up" task (example 2 in Chapter 3). This student has taken the three quantities given in the text and correctly performed two divisions, but these operations do not make any sense in this context (which is why a translation of the text is not necessary). Such a substitute strategy even seems to become more popular with age; in a school context, it makes sense

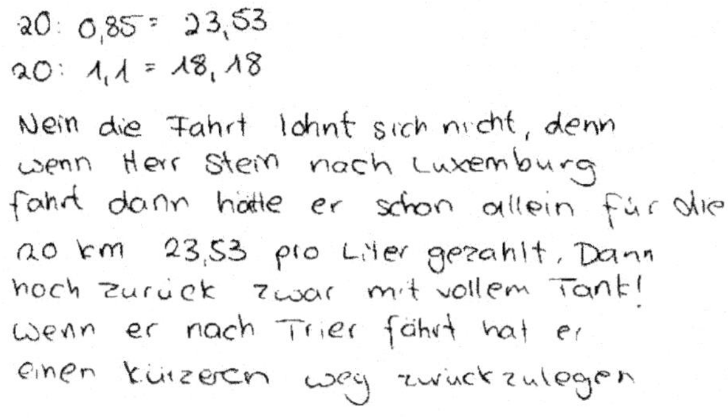

```
20 : 0,85 = 23,53
20 : 1,1 = 18,18
Nein die Fahrt lohnt sich nicht, denn
wenn Herr Stein nach Luxemburg
fahrt dann hätte er schon allein für die
20 km 23,53 pro Liter gezahlt. Dann
noch zurück zwar mit vollem Tank!
Wenn er nach Trier fahrt hat er
einen kürzeren Weg zurückzulegen
```

FIGURE 6.2 A student's solution to the "filling up" task

to follow it by attempting to please the teacher, especially to pass examinations, by at least doing something since the majority of contextual examination questions are often only dressed-up mathematics tasks that do not require an understanding of the given situation. This behaviour of students is empirically well documented in many countries around the world. Here is a well-known example from these empirical studies, the "Army Bus" task (Verschaffel et al., 2000): "450 soldiers must be bussed to their training site. Each army bus can hold 36 soldiers. How many busses are needed?" Popular answers according to the substitute strategy are "12 busses remainder 18" or "12.5 busses". Another example of a calculation that doesn't pay serious attention to the situation is this: "An orchestra needs 40 minutes to play Beethoven's 6th symphony. How long will it take to play Beethoven's 9th symphony?" (Hugh Burkhardt, personal communication). An answer sometimes heard is 60 minutes (the rule of three!), and a similar answer applies to "2 eggs need 6 minutes to get hard boiled. How long will 20 eggs need?" and also to "18 out of the 24 pupils in the class need 45 minutes for a written mathematics test, 6 are ill. How much time would all 24 students need?" The substitute strategy cannot be successfully applied with genuine modelling tasks since mostly it leads to an inadequate representation of the situation. Moreover, not all relevant information is always given, or sometimes the presentation of the task contains data that are irrelevant for the question to be answered. In other words, for genuine modelling tasks, it is indispensable to really understand the situation presented. The need to understand the situation in terms of the characteristic properties of the extra-mathematical context in which it is embedded is an essential reason why modelling tasks are difficult for students of all ages and backgrounds.

The *second step*, "simplifying and structuring the situation", is a source of difficulties as well. As Jankvist and Niss (2019) found in the abovementioned study, the predominant reason why several students gave inadequate solutions to modelling tasks was not that they did not know the mathematics needed or that they were unable to solve the resulting mathematical problems; they were unable to successfully pre-mathematise the situations in a way that was conducive to subsequent sensible mathematisation. The tasks in that study were deliberately chosen to

118 What we know from empirical research

require only basic mathematics; examples include the estimation of the height of a building on a photograph, the calculation of the average speed for a walk up and down a hill at two different speeds, and the determination of which of two circular pizzas of different sizes and prices gives the best value for money. The authors offer possible explanations of their observations. One is that some students – under the influence of the didactical contract they are accustomed to – did not pay serious attention to the formulation of the tasks, or to the conditions stated therein, but rushed off to find cues that may lead them to do some mathematical work they are familiar with. A second explanation is that some students did not reflect on the implications of the information provided in the task and failed to make relevant assumptions and specifications of the situation to be modelled. A third explanation is that some students were unable to undertake implemented anticipation of the steps needed for completion of pre-mathematisation and for arriving at a tractable mathematisation (cf. section 2.6).

Students must identify relevant variables in the given situation to find connections between these variables and to make assumptions concerning missing values of some variables. Making simplifying assumptions about real-world problems and clarifying the goals of the real model (and eventually selecting a model) were found by Frejd and Ärlebäck (2011) to be major stumbling blocks to Swedish upper secondary students getting started in the modelling process. In the "filling up" example (see Chapter 3), the identification of the two decisive variables, petrol consumption and tank volume of the car, requires students to take the situation seriously and to understand it in sufficient depth. In addition, they must make assumptions; not surprisingly, many students are afraid to make assumptions by themselves since, in many countries, they are not likely to have met such situations in school before. Rather, some students give up instead of making assumptions, as the following solution (Figure 6.3) to "filling up" in a test (within the DISUM project) shows:

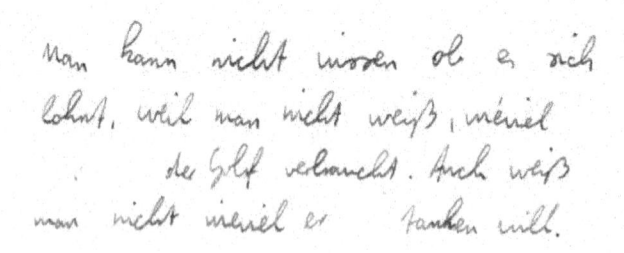

FIGURE 6.3 Another student's solution to the "filling up" task

("You cannot know if it is worthwhile since you don't know what the Golf [car] consumes. You also don't know how much she wants to fill up.") The student has clearly identified the two decisive variables for the basic model, petrol consumption and tank volume, which indicates that he has understood significant elements of the situation. However, he does not move on but stops his solution at this point, presumably because the values of these two quantities are missing in the text.

As indicated, the significance of implemented anticipation (see section 2.6), introduced theoretically by Niss (2010) and investigated empirically by Stillman and Brown (2014), Niss (2017), Werle de Almeida (2018) and Jankvist and Niss (2019), manifests itself during this second step because simplifying and structuring the situation, identifying relevant variables and relations between them, and making assumptions all have to be enacted with an eye to what could or should come next in the modelling process. That is, to make these choices and decisions, the modeller needs to project him- or herself into phases that have not yet occurred and to make use of that projection in the current phase.

The *third step*, "mathematising the simplified and idealised extra-mathematical (or real) model", often requires formalising verbal statements and writing down equations or functions to draw idealised geometrical configurations. This presupposes sufficient mathematical knowledge, including understanding the basic ideas behind the mathematical entities (the "Grundvorstellungen", see vom Hofe & Blum, 2016, which means the mental representations of these entities which carry their meaning). Undertaking mathematisation also requires devising a heuristic strategy, which is sometimes rather challenging (Stender, 2018; Albarracin & Gorgorió, 2014; Stillman et al., 2015; Galbraith et al., 2007). Once again, a crucial feature of the mathematisation step is the need for the student to perform implemented anticipation. An example of the need for basic ideas is seen in the "traffic flow" task (see example 6 in Chapter 3). Students must be able to condense the relation between the relevant variables speed, length, distance and flow to the equation $F = v/(l + d)$ where d itself depends on v. Experiences with students at all levels show that missing conceptual understanding of division leads to severe difficulties in establishing this equation since it is necessary to find a number (the number of cars on a road segment of a certain length) by dividing two lengths (the total length of the relevant street interval and the length that one car covers on the street). Positively speaking, a conceptual understanding of division as "sharing equally" will give immediate access to the term $v/(l + d)$ in concrete cases, for instance, for a 50 km street interval if the speed is 50 km/h, and the basic idea of a variable as "placeholder" admits a generalisation yielding the formula. More than that, the basic notion of function (either according to the conception of mapping or according to the conception of covariation: "What happens to y if x is changing?") is needed, too, since it is necessary to interpret this equation as a function $v \rightarrow F$ whose maximum leads to the solution of the problem. This is also a source of difficulties, for example, if a student cannot cope conceptually with composite functions or technically with rational functions. In several other examples in this book, functions emerge as natural models, for instance, linear functions in the taxi example (section 2.2), power functions in the loan amortisation example (example 3 in

Chapter 3) or exponential functions in the paper format example (example 4 in Chapter 3).

The *fourth step*, "working mathematically", is also a source of numerous challenges, difficulties and mistakes. The crucial element here is to devise a strategy to answer the mathematical questions that have arisen from the mathematisation step. Mastering the mathematical processes and reasoning involved in implementing the strategy once it has been devised also involves several different challenges. However, we abstain from dealing with such challenges, difficulties and mistakes here since these are part of mathematical problem solving at large and are not particular to modelling problems (see Schoenfeld, 1985, 1994; Lesh & Zawojewski, 2007).

The *fifth step*, "interpreting the mathematical results", is not equally demanding. In tasks set by the teacher, this step is sometimes forgotten since students think they are done after arriving at a mathematical result. However, because interpretation requires taking the extra-mathematical situation seriously and engaging mentally in it, there are also various possibilities of making mistakes resulting from a superficial engagement in the situation (also see Blum, 2011).

The *sixth step*, "validating and evaluating", often does not take place at all, unless the task has been devised by students themselves. It seems to be part of the "didactical contract" between teachers and students, in many parts of the world, that it is exclusively the teacher's responsibility to check the correctness and appropriateness of solutions. Here is an example: When asked to estimate the height of a five-storey building from a photograph which also includes a number of persons standing near the building, a student, after having used a scaling model based on the height of a door, concluded that the building was 1.6 km tall because he forgot to adjust the units (Jankvist & Niss, 2019). The student did not bother to check whether the result was reasonable; his job was, in his view, only to come up with an answer. And another example (for more details, see Blum & Borromeo Ferri, 2009): A group of students had correctly found, in the "lighthouse" task (see example 3 in Chapter 3), that a dot-like ship is 20 km away when it sees the lighthouse for the first time. Then they modified their model by assuming that the ship is 10 m high and found a distance of 16 km. Figure 6.4 shows the students' report ("From the horizon you can look ca. 20 km. . . . If the ship is 10 m high, the weather is good and the radius of the earth is 6370 km, then the lighthouse can be seen from ca. 16 km distance."). A validation of this solution would have made it obvious that something must be wrong since a taller ship will, of course, see the lighthouse earlier. We will come back to this example in section 6.4. Czocher (2018), in a small-scale study of engineering students, showed that validation and evaluation of model outcomes and models can be more complex than what is oftentimes thought and that it may be intrinsically intertwined with several other stages of the modelling process. As to the strategic aspects of evaluating the modelling process, focusing on the choice of ways in which such evaluation can be undertaken, also see Vorhölter (2018).

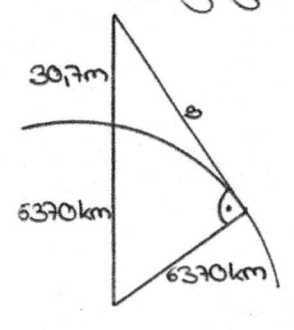

Annahmen
Erdradius: 6370 km
Wetterbedingungen: Kaltfront, kein Niederschlag

30,7 m

6370 km

6370 km

$s^2 = (6370 km + 30,7 m)^2 - (6370 km)^2$
$s^2 = (6370{,}0307 km)^2 - (6370 km)^2$
$s^2 = 391{,}1189 km^2$
$s = 19{,}78 km$
Vom Horizont aus sieht man ca. 20 km weit.
Neue Annahme: Schiff → 10 m hoch

$s^2 = (6370{,}0307 km)^2 - (6370{,}01 km)^2$
$s = 16{,}24 km$

Antwort:
Wenn das Schiff 10 m hoch ist, gute Wetterbedingungen herrschen und der Erdradius 6370 km beträgt, dann kann man den Leuchtturm aus ca. 16 km Entfernung sehen.

FIGURE 6.4 A students' solution to the "lighthouse" task

6.3 Individual modelling routes

We know from several studies (Matos & Carreira, 1997; Leiß, 2007; Borromeo Ferri, 2011; Schukajlow, 2011; Sol et al., 2011) that if students are dealing with modelling tasks independently, the modelling process is normally very irregular and does not follow those analytically constructed ideal-typical loops described in Chapter 2, representing a well-defined path, in any regular manner. Instead,

122 What we know from empirical research

actual modelling processes are characterised by jumps back and forth and in and out between the extra-mathematical world and mathematics, by the omission of certain steps or by "mini-loops". Borromeo Ferri (2007) speaks of "individual modelling routes", that is, individual sequences of steps within the complex scenario of connections between the extra-mathematical world and mathematics as depicted in various modelling cycles, influenced by individual knowledge, experiences and preferences such as individual thinking styles. To be more careful, one should speak of "visible individual modelling routes" since we only have access to externalisations of modelling processes (spoken or written texts, pictures, diagrams, formulae) and not to internal thought processes, even though we may have access to student conversations in small groups or are able to conduct interviews with students along the road or after the fact. Here is an example, referring again to the "lighthouse" task (see example 3 in Chapter 3). The students in a grade 10 class solved this task individually, without support (see Borromeo Ferri, 2007, for more information). Figure 6.5 shows the modelling routes taken by two students, embedded in the modelling cycle used in this study.

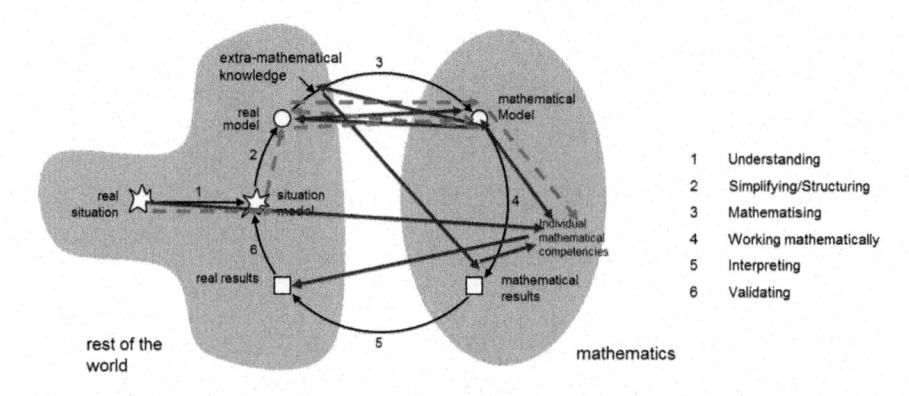

FIGURE 6.5 The modelling routes of two students

These two routes are rather distinct and seem to reflect the "thinking styles" of the two students (see Borromeo Ferri, 2007, for details).

6.4 Quality teaching of modelling

What do we know from empirical studies about effective teaching of modelling, related to the goals connected with modelling? Generally speaking, overarching findings about quality teaching of mathematics should also be valid for teaching mathematics in the context of relations to the extra-mathematical world or, in short, for teaching mathematical modelling, as long as these findings can be explained by certain theories about the learning of mathematics or human learning in general. Meta-studies such as Hattie (2009) give robust indications of which teaching approaches promise to cause learning effects and which do not. However, since teaching always takes place in environments that also depend on cultural factors, and since learning always depends (see section 6.1) on the specific context in which it takes place, we have to be careful about generalising specific empirical results. In the following, four important aspects of quality teaching (a–d) are discussed with particular regard to their role for teaching modelling.

a) Classroom management and learner orientation

A general result that is supported by many studies is that desirable learning effects can only be expected if the teacher maintains effective and learner-oriented classroom management (see, e.g., Hattie, 2009; Timperley, 2011; for a conceptual framework see Kunter & Voss, 2013). This includes structuring lessons clearly in continuation of a carefully thought-through didactico-pedagogical design of the lesson, with clear aims, using lesson time effectively, separating learning situations and assessment situations in a recognisable manner, linking the lesson(s) to students' existing knowledge, using students' mistakes constructively as learning opportunities, giving individual feedback and support, and varying methods and media flexibly. These are general aspects which are necessary but not sufficient for desirable learning effects. These aspects typically refer neither to the content nor to deeper cognitive structures of lessons but mostly to surface structures. They are, however, still relevant. Especially for modelling activities, group work seems particularly suitable (Ikeda & Stephens, 2001; Blomhøj & Kjeldsen, 2018). The group is not only meant to be a social but also, and especially, a cognitive environment (co-constructive group work; see Reusser, 2001). We also mention here the lesson plans described in Becker and Shimada (1997) which give a clear structure to lessons: First, the teacher sets the problem. Then, the students work in groups on the problem, supported by the teacher if needed. When the solutions are completed, some students present their solutions on the board, and afterwards these solutions are compared and discussed.

Part of effective classroom organisation is the appropriate use of media for explorations or presentations. In section 5.7, we discussed the possibilities, problems and implications of the use of digital tools. Among the media that may be of use in modelling environments are all media used in mathematics lessons in general, such as calculators, computers, projectors or drawing aids.

One important aspect of "learner-orientation" is to encourage students to find their own modelling pathways in dealing with modelling tasks, individually or in small groups. In everyday teaching practice, it is not unusual for teachers to strongly favour their own approach, often without even noticing it, possibly because of their limited knowledge of the "problem space" (Newell & Simon, 1971) or perhaps because they harbour definite beliefs about students' prerequisites. This often leads to a situation where, in the end, all students produce more or less the same approach and result, induced by the teacher's hints (Leikin & Levav-Waynberg,

2007; Borromeo Ferri & Blum, 2010b). Varying individual outcomes lead to multiple outcomes in the classroom. There are several reasons for encouraging multiple approaches and solutions to student tasks (Hiebert & Carpenter, 1992; Rittle-Johnson & Star, 2009; Tsamir et al., 2010; Brady, 2018). Thus, multiple approaches, when appropriate, reflect the genuine spirit of mathematical work and enable comparisons between, and meta-level reflections on, different approaches, even if the final results may be identical. It seems favourable to also encourage individual students to look for more than one approach. In the project MultiMa (Schukajlow & Krug, 2013; Schukajlow et al., 2015), two independency-oriented teaching units with modelling tasks were compared; in one unit, students were explicitly required to produce multiple approaches. Those students who developed several approaches had higher learning gains in a pretest-postest design with modelling tasks (mostly from the topic area of linear functions). Degrande et al. (2018) propose, as a result of an empirical investigation into fifth and sixth graders' solutions of word problems, encouraging multiple solutions from as early on as primary school.

b) Cognitive activation

It is essential to activate learners cognitively, that is, to stimulate students' own reflective activities, to induce them to establish their own mental structures. In particular, as "modelling is not a spectator sport" (Schoenfeld, personal communication), learning to model requires students to engage actively in modelling activities. This is not just a matter of general organisational structures such as whole-class teaching versus group work or individualised teaching, which are likely to depend on cultural backgrounds and traditions. The teaching methods must be chosen to ensure that learners become and remain cognitively active (Schoenfeld, 1994). One way to stimulate students' mental activities is to let them work as independently as possible, modulo the didactico-pedagogical aims of the activities. Here, one should distinguish carefully between students working independently but with teacher support and students working alone, totally on their own. Many studies (for instance, DISUM, see below) show that working alone may result in students getting hopelessly stuck and learning nothing. What is required, instead, according to many studies (see, e.g., Blum, 2011; Stender & Kaiser, 2016), is a continual balance between students' independence and teacher's guidance, in accordance with Aebli's "Principle of minimum support" (Aebli, 1985), which can be perceived as an amendment to Vygotsky's "Zone of Proximal Development". This calls for adaptive teacher interventions (Leiß, 2010), which means interventions which are as limited as possible and allow students, in case of difficulties and barriers, to continue their work as independently as possible. Whether an intervention is effective can – in principle – only be judged afterwards: Has the cognitive barrier really been overcome? Adaptive interventions can be regarded as a special case of "scaffolding" (see Bakker et al., 2015). In everyday classrooms, teachers often resort to interventions focused on mathematical content, sometimes to prevent mistakes or blockages from occurring. According to several studies (see, e.g., Leiß, 2007), teachers use strategic interventions sparingly, and most interventions are not adaptive. Strategic interventions (examples below) focus on the process at a meta-level and not on the content. They have, by their nature, the potential of being particularly adaptive. Generally speaking, the nature and extent of teachers' interventions in students' work is largely shaped by the didactical contract in the sense of Brousseau (1997) as well as by the teachers' devolution of tasks to students.

A manifest example of a successful strategic intervention was observed in the context of the "lighthouse" task (example 3 in Chapter 3) in a German grade 10 class (for more details, see Blum & Borromeo Ferri, 2009). Two groups of students had created a wrong model (shown in the end of section 6.2) by simply inserting the height of the ship into the Pythagorean formula instead of placing the ship beyond the horizon and drawing a second right-angled triangle. The other groups had not taken the height of the ship into account. The teacher noticed this mistake while observing the students' work but refrained from intervening. Instead, he checked the implications of the mistake by calculating for himself further distances for various ship heights (1 m, 5 m, 10 m, . . .). He then let one group present its work to the whole class, and only after all students' presentations (where no student discovered the mistake) did he produce a cognitive conflict by showing his calculations for various ship heights and saying (translated from German):

> I would like you to try to imagine the relationship between the height of the ship and the distance between lighthouse and ship once again by means of a sketch. Is the way in which this calculation was done really correct? I have just done calculations exactly in the way Max presented it here.

The students in one of the "erring" groups then found their mistake, corrected their model based on an appropriate sketch that they developed independently after the teacher's stimulus and eventually presented the corrected model to the class.

The following list offers some examples of strategic interventions in modelling environments that proved supportive in the DISUM project (see Schukajlow et al., 2012):

1. Read the text carefully!
2. Imagine the real situation clearly!
3. Make a sketch!
4. What do you aim at?
5. What is missing?
6. Which data do you need?
7. Have you already dealt with a similar problem?
8. How far have you got?
9. Does this result make sense for the real situation?

Concerning the intervention "Make a sketch", Rellensmann et al. (2017) found, in a study with 61 ninth graders, positive effects of the prompt to produce a drawing on the students' modelling performance in a test with tasks related to the Pythagorean theorem. More precisely, the students were asked to produce, for each task, both a "situational drawing" capturing the fundamentals of the real-world situation and a "mathematical drawing" illustrating the mathematical model. The accuracy of the mathematical drawings correlated strongly with the modelling performance while the influence of the quality of the situational drawing on the quality of the mathematical drawing was mediated by the students' strategic knowledge on drawings. An in-depth analysis of students' work showed that the prompt to make a situational drawing helped students understand the situation given, the first step in the solution process (see section 6.2). A possible consequence for the teaching of modelling is to foster and further students' knowledge of drawings, as part of developing their meta-knowledge, and, when appropriate, advise them to make a situational as well as a mathematical drawing.

Stender (2018) shows how heuristic strategies in the sense of Polya (1973), such as "Break down the problem into sub-problems", "Explore extreme cases" or "Combine special cases to the general case", may lead directly to corresponding strategic teacher interventions. These interventions proved successful in a study with the same task, which means they helped most students to overcome their blockages and to continue with their independent work.

c) Meta-cognitive activation

Another important criterion for quality teaching is the activation of learners not only cognitively but also meta-cognitively (see Schoenfeld, 1987, for the role of meta-cognition in mathematical learning). All activities should be accompanied by reflections on the fly and, in retrospect, with the aim to advance appropriate learning and problem-solving strategies, cognitive as well as meta-cognitive, such as planning, monitoring, regulating or evaluating. Above, we mentioned promising strategies such as "Make a sketch" or "Decompose the problem into sub-problems". There are empirical results concerning positive effects of using strategies for modelling activities (Tanner & Jones, 1995; Schoenfeld, 1994; Matos & Carreira, 1997; Stillman & Galbraith 1998; Goos, 2002; Kramarski et al., 2002; Stillman, 2011; Schukajlow & Krug, 2013; for an overview, see Greer & Verschaffel, 2007; Vorhölter et al., 2019). Recently, Vorhölter (2018) proposed a way of conceptualising and measuring meta-cognitive modelling competencies and utilised it with 431 grade 9 students to identify the meta-cognitive strategies they used. She found that three sorts of strategies prevailed: strategies meant to ensure a smooth modelling process; strategies used for regulation when problems occur; and strategies for evaluating the entire modelling process.

An example of a strategic instrument for modelling is the "Solution Plan" which was used in the DISUM project for the lower secondary level. This is a four-step schema summarising how modelling tasks can be solved ("Understanding task/ Searching mathematics/Using mathematics/Explaining result"; see Figure 6.6 and see Blum, 2011, for more details).

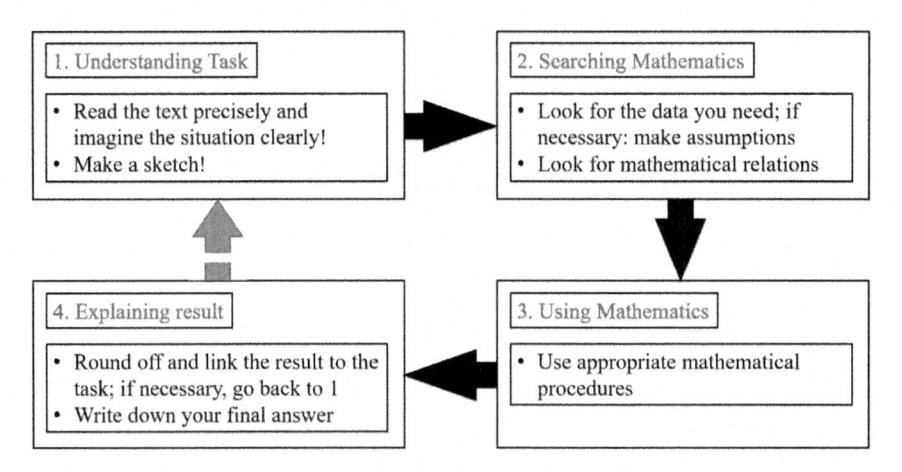

```
┌─────────────────────────────────┐         ┌─────────────────────────────────┐
│ 1. Understanding Task           │         │ 2. Searching Mathematics        │
├─────────────────────────────────┤   ───▶  ├─────────────────────────────────┤
│ • Read the text precisely and   │         │ • Look for the data you need; if│
│   imagine the situation clearly!│         │   necessary: make assumptions   │
│ • Make a sketch!                │         │ • Look for mathematical relations│
└─────────────────────────────────┘         └─────────────────────────────────┘
              ▲                                            │
              │                                            ▼
┌─────────────────────────────────┐         ┌─────────────────────────────────┐
│ 4. Explaining result            │         │ 3. Using Mathematics            │
├─────────────────────────────────┤   ◀───  ├─────────────────────────────────┤
│ • Round off and link the result │         │ • Use appropriate mathematical  │
│   to the task; if necessary,    │         │   procedures                    │
│   go back to 1                  │         │                                 │
│ • Write down your final answer  │         │                                 │
└─────────────────────────────────┘         └─────────────────────────────────┘
```

FIGURE 6.6 The DISUM "Solution Plan" for modelling tasks

As is the case with most versions of the modelling cycle, this is not meant to be a schema that students must follow but a guideline, particularly useful in case difficulties arise. It turned out that ninth graders who used this instrument had higher learning gains in a test with modelling tasks after a five-lesson teaching unit on modelling than did students to whom it was not available (for details, see Schukajlow et al., 2015).

d) Challenging content

Seen from the perspective of mathematics as a subject, one of the most important quality criteria is challenging orchestration of the content, especially by means of substantive tasks. Students need continual opportunities to develop and practice the aspired competencies at varying cognitive levels. Considering our knowledge about situated cognition (see section 6.1), it is vital that students have extensive opportunities to deal with various kinds of modelling tasks, with varying extra-mathematical contexts and varying mathematical content. As stated in section 5.6, authenticity of the contexts is not always required. However, empirical studies show that if contexts are more authentic, the "suspension of sense-making" (see section 6.2) can be reduced considerably (Palm, 2007; Verschaffel et al., 2010; Kaiser & Schwarz, 2010). For instance, if the "Army bus" task (section 6.2) is embedded in a credible context in which students have to fill in an order form to a bus company, the number of reasonable solutions increases considerably. Empirical studies such as Carreira and Baoia (2018) show that students are willing to engage also in non-authentic contexts if the contexts are credible to them; the context presented in this study of ninth graders is the mixing of blue, yellow and white paint to obtain desired shades of green.

Substantive tasks as a key component in quality teaching are not only a matter of single lessons or units but also must be kept in mind for long-term planning of teaching. Competencies develop in long-term learning processes, and learning is always "further learning" by linking new content to learners' pre-knowledge. An important aspect of quality teaching is to develop well-coordinated sequences of units and tasks that enable long-term competency development, in particular of modelling competency/ies; see section 6.7 for more reflections on competency development.

6.5 A "successful" project with a "successful" lesson

Several studies (both case studies and intervention studies) have shown that mathematical modelling can be learned by secondary school and tertiary students provided they receive quality teaching (see, e.g., Kaiser-Meßmer, 1987; Galbraith & Clathworthy, 1990; Abrantes, 1993; Maaß, 2007; Biccard & Wessels, 2011; Schukajlow et al., 2012; Blomhøj & Kjeldsen, 2018). Some studies have demonstrated that students' beliefs about mathematics can be broadened by way of quality teaching. We refer here to the DISUM project (for more details, see Blum & Leiß, 2007; Blum, 2011; Schukajlow et al., 2012).

In the main part of the DISUM project (2009–2013), two teaching units of 10 lessons each for grade 9 on modelling were conceived in two versions: one based on an "operative-strategic" teaching design and one making use of a "directive" teaching design. The guiding principles of the "operative-strategic" design were as follows:

1 Teacher's guidance aiming at students' active and independent modelling processes, with adaptive interventions;
2 Change between independent work in small groups and whole-class activities (concerning students' presentations and retrospective reflections); and
3 Teacher's coaching based on the "Solution Plan" mentioned in section 6.4.

The guiding principles of the "directive" design were as follows:

1 Development of common solution patterns by the teacher; and
2 Change between whole-class teaching (oriented towards the "average student") and students' individual work in exercises.

Both "operative-strategic" and "directive" teaching styles were conceived as idealtypical styles for independence-oriented respectively teacher-directed teaching, realised by experienced teachers from a preceding reform project ("SINUS", see Blum, 2008) who were particularly trained for the project. Both styles and each single lesson were described in detail in two so-called prompt books which formed the basis for the training. The modelling tasks were identical in both designs and were treated in the same order. In the first round of the project, 7 + 7 grade 9 classes from the intermediate level ("Realschulen") took part. All classes were reduced to 16 students each, by means of a preceding mathematics achievement test, to have comparable groups in terms of mathematical performance. The teaching unit was framed by a pre-test immediately before lesson 1 and a post-test immediately after lesson 10, consisting both of modelling tasks and intra-mathematical tasks in the content areas of linear functions and the Pythagorean theorem. The test results showed a significantly higher learning progress of the "operative-strategic" classes, stemming from their bigger progress in modelling, whereas the learning progress concerning the intra-mathematical skills and abilities were identical. The "directive" classes had no significant learning gains in modelling at all – a remarkable result. From a normative point of view, the progress of half a standard deviation was not satisfactory, even for the "operative-strategic" classes. In the second round of the project, one year later, an "integrated" teaching design was conceived which contained both operative-strategic and directive elements and in which both the students and teachers had access to the Solution Plan. The "directive" elements integrated into the "operative-strategic" design were those which, according to classroom observations and theoretical considerations, had the biggest potential to improve the "operative-strategic" design, namely an introduction of the Solution Plan by a teacher demonstration of an ideal-typical handling of a modelling task (using the teacher as a "model") and individual practising in the last two lessons of the unit. The progress in the re-designed 10-lesson unit was now markedly bigger, nearly one standard deviation.

We present here a typical lesson from the integrated design. The modelling task treated in this lesson is "filling up" (see example 2 in Chapter 3).

1 The task is distributed to the class, and a student reads it aloud; only questions concerning the text are allowed ("What kind of a Golf does Mrs. Stein have?"), no hints for solutions are given; each student has, on his/her desk, a "working sheet" with guidelines concerning the general approach to dealing with the task ("first read the text and work by yourself for a while, then exchange your approaches and results within your group, . . .") as well as the four-step Solution Plan (which had been introduced in the third lesson of the unit).
2 Each student works by him/herself on the task; the teacher walks around, observes, diagnoses, encourages and gives minimal support if necessary (such as "Imagine the situation concretely; accompany Mrs. Stein in your mind and think about what she could be doing" or "Make a sketch") or responds to information questions (such as "How much petrol does such a Golf consume?").
3 The students sit in groups of three to four students in the beginning; after a while, the individual approaches are compared within each group; the teacher continues to observe and to give minimal guidance.
4 Each student completes his/her modelling work individually; the teacher encourages activities to validate the model results.

5 Some solutions (models and results) are presented to the whole class (some of the presenters have been identified and asked by the teacher before, and some students volunteer spontaneously).
6 The solutions are compared and discussed (especially numerical approaches with differing assumptions and, if available, algebraic approaches), and further variables such as time and risk are put forward, provoked by the teacher.
7 The modelling processes are analysed in retrospect, aided by the four-step Solution Plan.
8 The students get the opportunity to improve their individual approaches and outcomes, supported by their neighbours; the teacher requests that no student should copy one of the models presented but seek to improve his or her own work.
9 Homework: Each student writes down his/her "optimal" individual solution.

Whether this lesson satisfies the quality criteria described in section 6.4 depends not only on the global structure of the lesson and on the underlying prompt book but equally on the teacher's concrete actions in the classroom, which can only be judged by analysing the actual lesson in detail. However, several elements of quality teaching can be identified in this lesson description, such as a variation of methods to activate the learners cognitively and to encourage and support their independent modelling, encouragement of multiple approaches, retrospective reflections and thus a fostering of strategies, and a manifest cognitive demand of the task given. Classroom observations, video analyses of the lessons and student questionnaires (administered in each lesson) indicate that the treatment was implemented appropriately and the students had ample opportunities to model independently.

6.6 Teacher competencies for teaching modelling

From several studies, we know that, apart from the students and their attitudinal and affective approach to education and the general structural and cultural environment of schools and schooling, the most important "variable" in the teaching-learning situation is the teacher. His/her professional competencies have a paramount influence on the quality of what happens in the classroom and on students' learning progress. This especially holds for teaching mathematical modelling (Stender et al., 2017; Maaß & Engeln, 2018).

In terms of general teaching quality in mathematics, results from the COACTIV project (see Kunter et al., 2013) show an immense influence of the teachers' pedagogical content knowledge (PCK) in mathematics on the quality of instruction and on the students' mathematical achievement. This project was conceptually and technically embedded in the German PISA project, especially in the longitudinal study of 2003–2004 where whole ninth grade classes were tested. The teachers of these classes were representative of secondary mathematics teachers in Germany. In COACTIV, these teachers were extensively tested and examined so that the data of the teachers and their students could be considered together. In particular, the teachers' PCK was conceptualised and tested in three dimensions: the ability to find several solutions to PISA mathematics tasks; the ability to explain certain mathematical content in multiple ways (such as why "minus times minus yields plus"); and the ability to predict mistakes based on diagnoses of similar ones. It turned out that the most important variables, mediating between teachers' competencies and students' learning by influencing the quality of instruction (see section 6.4), were classroom management; teachers' constructive support for learners; and the cognitive levels of the mathematical tasks used on class tests. The teachers' mathematical content knowledge (CK) was not directly related to those mediating variables but proved to be an important source for the PCK.

Modelling was not the main focus of the COACTIV study, although the teacher tests also contained tasks related to modelling (for instance, writing down multiple solutions for an everyday problem involving proportional functions). What competencies a teacher needs to devise and carry out high-quality teaching activities for modelling is still an open research question. This is not surprising given the immense variation of economic, cultural, structural, organisational and traditional boundary conditions for teaching in different parts of the world. On a theoretical level, the four-dimensional model shown in Figure 6.7 (see Borromeo Ferri & Blum, 2010a; Borromeo Ferri, 2017) comprises important modelling-related components of teachers' PCK.

Theoretical dimension	a) Modelling cycles b) Aims & perspectives of modelling c) Types of modelling tasks
Task dimension	a) Multiple solution of modelling tasks b) Cognitive analyses of modelling tasks c) Construction of modelling tasks
Instruction dimension	a) Planning lessons with modelling tasks b) Carrying out lessons with modelling tasks c) Interventions, support and feedback
Diagnostic dimension	a) Recognising phases in modelling process b) Recognising difficulties and mistakes c) Marking tests

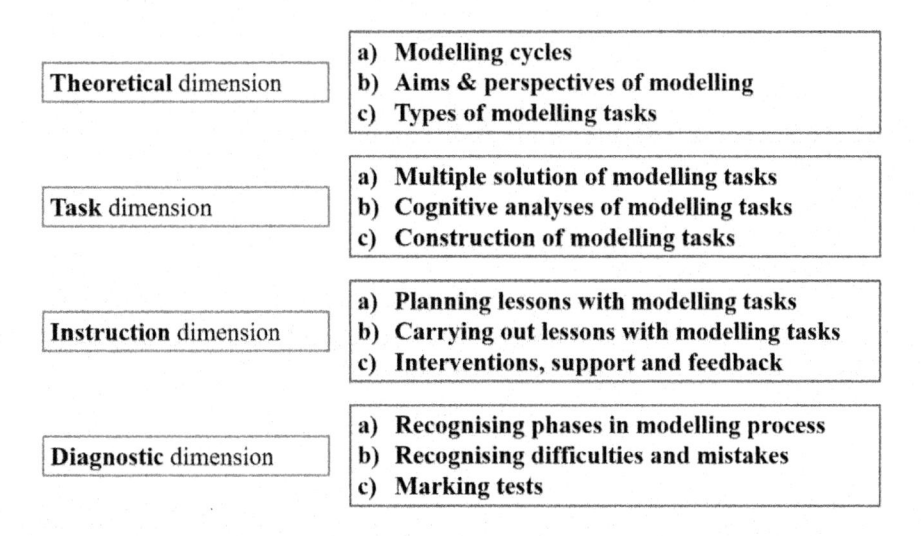

FIGURE 6.7 Model of teacher competencies for modelling
Source: Borromeo Ferri, 2017

Most of these aspects are addressed in this book. Some aspects, such as carrying out lessons, can only be advanced by work in practice. It is a major task of teacher education, both pre-service and in-service, to advance those competencies. Borromeo Ferri (2017) describes an evaluated module for implementing mathematical modelling into teacher education, oriented towards the abovementioned competency model. This module starts with solving and developing modelling tasks for school, accompanied by theoretical reflections on modelling. After analysing school students' solutions to some of these tasks as well as video clips from modelling lessons, using criteria for quality teaching, teacher students carry out a modelling unit by themselves. Afterwards, they report about their experiences. For more details of this module, see Borromeo Ferri (2017, pp. 3–12). Criteria of effective in-service teacher education, not only concerning mathematical modelling, are described in section 5.3.

6.7 Modelling competencies

Along with the manifest growth of interest in modelling competencies, research began to empirically study students' modelling competency/ies, including the development of such competencies over time. In the following section, we will mention two examples.

Frejd and Ärlebäck (2011) charted the modelling competencies among Swedish upper secondary students and found that students' modelling competencies were only weakly developed and that making simplifying assumptions about real-world situations constituted the most significant challenge to students. In a comparative study of 1108 upper secondary German and Chinese students' modelling competencies, Ludwig and Xu (2010) found that while there were no gender differences among German students, there were for the Chinese students (girls performed better than boys). The study further found that in both countries, students increased their modelling competencies with age/level, and the progress was faster among Chinese students in the upper grades.

Competencies such as the modelling competency ideally are developed in long-term learning processes, beginning in primary school with "implicit models" (Greer & Verschaffel, 2007; Borromeo Ferri & Lesh, 2013) and continuing over the years. We know from research that a necessary means for such a long-term competency development is sustained and intelligent practising of mathematical modelling activities. "Intelligent" means that all aspects (sub-competencies, content, context, cognitive demand) are deliberately varied and linked by the teacher, both within and across teaching units, and that learners are made conscious of these variations and connections. Some studies (see Blomhøj & Jensen 2003; Kaiser & Brand, 2015) indicate that both purely "atomistic" approaches, where the emphasis is on the development of modelling sub-competencies, and purely "holistic" approaches, where the emphasis is on the modelling competency as a whole (see section 3.8), have their shortcomings. A balance between sub-competencies and the modelling competency as a whole is advisable. It is an open question as to what such a balance should look like to maximise desirable learning outcomes.

What we need, both for theoretical and practical reasons, is a competency development model for modelling which is theoretically sound and empirically well founded. One approach to characterising an individual's competency development over time comes from the Danish KOM project (Niss & Jensen, 2002; Niss & Højgaard 2011; Blomhøj & Jensen, 2007). The authors distinguish between three dimensions in an individual's possession of a given mathematical competency (cf. Chapter 2): the "degree of coverage" of the defining aspects of this competency; the "radius of action" indicating the spectrum of contexts and situations within which the individual can activate the competency; and the "technical level" which indicates the conceptual and technical level of the mathematical entities that the individual can bring to modelling situations. Progress in the development of the modelling competency means progress in at least one of these dimensions and decrease in none of them.

6.8 Assessment of mathematical modelling

Assessment of students' mathematical modelling attracted the interest of researchers in the field at a rather early stage of its development (e.g., Gillespie et al., 1989; Francis & Hobbs, 1991). Thus, Niss (1993) identified what he saw as the main challenge for this endeavour: "How to shape and practice assessment in the area of application and modelling in such a way that assessment serves its purpose without destroying the application and modelling work?" (p. 44). This is in line with the general observation that assessment often becomes so simplistic and superficial that it reduces, compromises or even distorts the very essence of what it purports to assess. In this section, we concentrate on summative assessment and abstain from discussing modes of formative assessment accompanying learning processes.

Initially, the emphasis was on devising schemes for the assessment of students' modelling work as reflected in written reports of their work, typically giving rise to rating scales (Haines, 1991; Naylor, 1991; Money & Stephens, 1993; Haines & Izard, 1995; Ikeda & Stephens, 1998; Haines et al., 2001). Such assessment focuses on students' products, that is, the modelling work they carried out and their reports thereof. Only indirectly is this an assessment of students about their individual skills and competencies as modellers. For such assessment to be possible, it is necessary not only to consider a wide range of products produced by the individual student but also to capture the processes, deliberations, strategies, choices, decisions and arguments which the student activates in his or her modelling work. This requires access to more than the final modelling products, typically through observations of the individual student at work or interviews conducted along the road or after the completion of a number of modelling tasks. This led to a growing interest in creating ways to assess students' modelling skills (Haines et al., 1993; Haines et al., 2000, 2001), including progress in the development of such skills (Izard, 2007), an endeavour which soon after was transposed into attempts to assess students' modelling competency or (sub-)competencies (see Chapter 4), e.g., by Jensen (2007), Haines and Crouch (2007), Henning and Keune (2007) or Zöttl et al. (2011). As an example of a criteria-based assessment scheme, we mention the one developed by Dunne and Galbraith (2003, p. 19) in which modelling is being assessed in terms of four criteria, each of which can be accomplished at three different levels: C (lowest),

B (medium) and A (highest). The criteria are "ability to specify problem clearly"; "ability to set up model"; "ability to solve, interpret, validate, refine"; and "ability to communicate results".

It is characteristic that most modes and instruments for the assessment of modelling products as well as individual students' modelling competency/(sub-)competencies (such as the Dunne and Galbraith scheme just described) are tightly connected to some version of the modelling cycle (see Chapter 2). This is hardly surprising since full-scale modelling and the ability to carry it out consists of handling the entire modelling process in all its facets. This has given rise to the question of how to do this. Basically, there are two kinds of approaches to assessing modelling products and modelling competency/cies: holistic approaches and atomistic approaches (also see Houston, 2007). Holistic approaches seek to assess the entire modelling work or process at the same time in "one shot", that is, all the aspects of modelling activity: identifying and posing model generating questions pertaining to some extra-mathematical context and situation; undertaking pre-mathematisation (including making assumptions and simplifications, procuring information and collecting data); undertaking mathematisation by translating extra-mathematical questions and entities into mathematical questions and entities referring to some chosen mathematical domain; working mathematically with and within the model established to derive mathematical answers to the mathematical questions posed; de-mathematising the mathematical outcomes (i.e., interpreting the mathematical answers in terms of the extra-mathematical context and situation at issue); validating the model outcomes in terms of their validity and relevance for the situation modelled; and evaluating the model in terms of its intrinsic and extrinsic quality as a model, possibly vis-à-vis alternative models available. Atomistic approaches to the assessment of mathematical modelling refrain from taking all aspects of modelling work and processes into account at the same time but seek to assess only one or a few aspects at a time, typically by zooming in on one or two steps in the ideal-typical modelling cycle. For an atomistic approach to provide comprehensive assessment of a student's modelling competency/cies, it is necessary to collect and combine focused assessments of sufficiently many and sufficiently varied different aspects and steps in the modelling process as undertaken by the student being assessed.

This points to the need for identifying, designing and investigating possible modes of assessment, suitable for holistic or atomistic approaches to the assessment of mathematical modelling, respectively. Frejd, in his extensive literature review (Frejd, 2013) of 76 articles dealing with assessment of mathematical modelling, pays particular attention to such modes of assessment. His study identified five major modes of assessment adopted in the literature to assess mathematical modelling: written tests, projects, hands-on tests, portfolios, and contests, although the employment of these modes is not restricted to the assessment of modelling. The choice of a suitable assessment mode is linked to the aim of the assessment objective at issue. Written tests – including the tests used in large-scale international assessment programmes such as PISA (Turner, 2007) – are predominantly used in atomistic approaches, whereas projects and portfolios, and in some contexts contests as well, are favoured in holistic approaches. The abovementioned assessment scheme developed by Dunne and Galbraith is an instance of a holistic approach. The same is true of the project-based modelling exams in Denmark as described by Antonius (2007) and of the laboratory-like assessment in the Dutch TIMSS supplement as described by Vos (2007). Recent contributions such as the discussion in Djepaxhija et al. (2017) about the suitability of multiple choice tasks for assessing modelling competencies show how topical the questions related to assessment continue to be for the modelling discussion.

The assessment issues mentioned have primarily dealt with the assessment of modelling products and with students' learning of mathematical modelling as reflected in their modelling competencies. However, occasionally researchers have also investigated assessment as to which specimen of teaching is conducive to the fostering and furthering of students' modelling competency (for an example, see Izard et al., 2003), even though we are not in a position to claim the existence of a larger bulk of systematic research with this focus.

6.9 Models as a vehicle in learning mathematics: the models and modelling perspective

In section 2.8, we introduced the non-contradictory distinction between two overarching purposes of mathematical modelling in mathematics education (see also Julie & Mudaly, 2007, for a somewhat similar but not completely identical distinction). The first purpose sees modelling as an independent goal of mathematics teaching and learning, whereas the second sees modelling as a vehicle for something else, above all the learning of mathematics as a subject.

While the bulk of this book focuses on modelling as an independent goal, the present section focuses on empirical research pertaining to the second purpose: modelling as a vehicle for the learning of mathematics. Several studies (see the comprehensive survey in Schukajlow et al., 2018) have shown that modelling can help foster motivation, interest and sense-making with regard to the learning of mathematical concepts. Among others, Ottesen (2001) and Blomhøj and Kjeldsen (2010) have shown how modelling can facilitate the learning of tertiary mathematics.

A particularly broad approach to utilising mathematical models and mathematical modelling as a vehicle for students' own construction and consolidation of mathematical concepts is the so-called Models and Modelling Perspective (compare sections 2.8 and 4.6), first introduced by Lesh and Doerr (2003). In this approach, the notion of Model Eliciting Activities (MEAs) as well as their extension to larger sequences, called Model Development Activities (MDAs), are crucial components. In these activities, students are presented with an extra-mathematical situation and are invited to make sense of it by (re)inventing or (re)creating mathematical concepts – for example, ratio, scale, average, spread, rate of change, function, etc. – to represent key aspects of the situation given. Students' experiences gained from this kind of representational and sense-making work will not only help consolidate the mathematical concepts thus established but also serve the purpose of making these concepts available for dealing with future extra-mathematical situations to which they are actually or potentially relevant. Lesh and Harel (2003), for instance, focused on the ways in which proportional reasoning could be consolidated by MEAs in the early grades. Lesh et al. (2000) presented an MEA in which middle school students were given measurement data of a number of aspects of the flight performance of six different paper aeroplanes and were asked to come up with an assessment of the paper planes with respect to four different sorts of flight characteristics: the best floater; the most accurate plane; the best boomerang; and the best overall paper plane. Ärlebäck et al. (2013) and Ärlebäck and Doerr (2018) have demonstrated how MDAs were used to underpin prospective engineering students' understanding rates of change pertaining to a range of phenomena in physics. To mention a final example: Brady (2018) showed how MEAs and MDAs could support in-service mathematics teachers' imaginative engagement with graphical representations of linear function models.

References

Abrantes, P. (1993). Project work in school mathematics. In: J. DeLange (Ed.), *Innovation in Maths Education by Modelling and Applications* (pp. 355–364). Chichester: Horwood.
Aebli, H. (1985). *Zwölf Grundformen des Lehrens*. Stuttgart: Klett-Cotta.
Albarracin, L. & Gorgorió, N. (2014). Devising a plan to solve Fermi problems involving large numbers. In: *Educational Studies in Mathematics* **86**, 79–96.
Anderson, J.R., Reder, L.A. & Simon, H.A. (1996). Situated learning and education. In: *Educational Researcher* **25**(4), 5–11.
Antonius, S. (2007). Modelling based project examination. In: W. Blum, P.L. Galbraith, H-W. Henn & M. Niss (Eds.), *Modelling and Applications in Mathematics Education: The 14th ICMI Study* (pp. 409–416). New York, NY: Springer.
Ärlebäck, J. & Doerr, H. (2018). Students' interpretations and reasoning about phenomena with negative rates of change throughout a model development sequence. In: *ZDM: The International Journal on Mathematics Education* **50**(1 + 2), 187–200.
Ärlebäck, J., Doerr, H. & O'Neill, A. (2013). A modeling perspective on interpreting rates of change in context. In: *Mathematical Thinking and Learning* **15**(4), 314–336.
Bakker, A., Smit, J. & Wegerif, R. (2015). Scaffolding and dialogic teaching in mathematics education: Introduction and review. In: *ZDM: The International Journal on Mathematics Education* **47**(7), 1047–1065.
Baruk, S. (1985). *L'age du capitaine. De l'erreur en mathematiques*. Paris: Seuil.
Becker, J.B. & Shimada, S. (Eds.) (1997). *The open-ended approach: A new proposal for teaching mathematics*. Reston: NCTM.
Biccard, P. & Wessels, D.C.J. (2011). Documenting the development of modelling competencies of grade 7 students. In: G. Kaiser, W. Blum, R. Borromeo Ferri & G. Stillman (Eds.), *Trends in Teaching and Learning of Mathematical Modelling (ICTMA 14)* (pp. 375–383). Dordrecht: Springer.
Blomhøj, M. & Jensen, T.H. (2003). Developing mathematical modelling competence: Conceptual clarification and educational planning. In: *Teaching Mathematics and its Applications* **22**(3), 123–139.
Blomhøj, M. & Jensen, T.H. (2007). What's all the fuss about competencies? In: W. Blum, P.L. Galbraith, H.-W. Henn & M. Niss (Eds.), *Modelling and Applications in Mathematics Education* (pp. 45–56). New York: Springer.

Blomhøj, M. & Kjeldsen, T.H. (2010). Learning mathematics through modelling: The case of the integral concept. In: B. Sriraman, L. Haapasalo, B.D. Søndergaard, G. Palsdottir & S. Goodchild (Eds.), *The First Sourcebook on Nordic Research in Mathematics Education* (pp. 569–581). Charlotte, NC: Information Age Publishing.

Blomhøj, M. & Kjeldsen, T.H. (2018). Interdisciplinary problem oriented project work: A learning environment for mathematical modelling. In: S. Schukajlow & W. Blum (Eds.), *Evaluierte Lernumgebungen zum Modellieren* (pp. 11–29). Wiesbaden: Springer Spektrum.

Blum, W. (2008). Opportunities and problems for "quality mathematics teaching": The SINUS and DISUM projects. In: M. Niss (Ed.), *ICME-10 Proceedings, Regular Lectures*. Roskilde: IMFUFA.

Blum, W. (2011). Can modelling be taught and learnt? Some answers from empirical research. In: G. Kaiser, W. Blum, R. Borromeo Ferri & G. Stillman (Eds.), *Trends in Teaching and Learning of Mathematical Modelling (ICTMA 14)* (pp. 15–30). Dordrecht: Springer.

Blum, W. & Borromeo Ferri, R. (2009). Mathematical modelling: Can it be taught and learnt? In: *Journal of Mathematical Modelling and Application* 1(1), 45–58.

Blum, W. & Leiß, D. (2007). Investigating quality mathematics teaching: The DISUM Project. In: C. Bergsten & B. Grevholm (Eds.), *Developing and Researching Quality in Mathematics Teaching and Learning, Proceedings of MADIF 5* (pp. 3–16). Linköping: SMDF.

Boaler, J. (1993). Encouraging the transfer of "school" mathematics to the "real world" through the integration of process and content, context and culture. In: *Educational Studies in Mathematics* **25**, 341–373.

Borromeo Ferri, R. (2007). Modelling problems from a cognitive perspective. In: C.R. Haines, P.L. Galbraith, W. Blum & S. Khan (Eds.), *Mathematical Modelling: Education, Engineering and Economics* (pp. 260–270). Chichester: Horwood.

Borromeo Ferri, R. (2011). *Wege zur Innenwelt des mathematischen Modellierens: Kognitive Analysen zu Modellierungsprozessen im Mathematikunterricht*. Wiesbaden: Vieweg+Teubner.

Borromeo Ferri, R. (2017). *Learning How to Teach Mathematical Modeling in School and Teacher Education*. Cham: Springer.

Borromeo Ferri, R. & Blum, W. (2010a). Mathematical modelling in teacher education: Experiences from a modelling seminar. In: V. Durand-Guerrier, S. Soury-Lavergne & F. Arzarello (Eds.), *CERME-6: Proceedings of the Sixth Congress of the European Society for Research in Mathematics Education* (pp. 2046–2055). Lyon: INRP.

Borromeo Ferri, R. & Blum, W. (2010b). Insight into teachers' unconscious behaviour in modeling contexts. In: R. Lesh, C.R. Haines, P.L. Galbraith & A. Hurford (Eds.), *Modeling students' mathematical modeling competencies* (pp. 423–432). New York: Springer.

Borromeo Ferri, R. & Lesh, R. (2013). Should interpretation systems be considered to be models if they only function implicitly? In: G. Stillman, G. Kaiser, W. Blum & J. Brown (Eds.), *Teaching Mathematical Modelling: Connecting to Teaching and Research Practice: The Impact of Globalisation* (pp. 57–66). New York: Springer.

Brady, C. (2018). Modelling and the representational imagination. In: *ZDM: The International Journal on Mathematics Education* **50**(1 + 2), 45–59.

Brown, J.S. (2019). Real world task context: Meanings and roles. In: G. Stillman & J. Brown (Eds.), *Lines of Inquiry of Mathematical Modelling Research in Education* (pp. 53–81). Cham: Springer.

Brousseau, G. (1997). *Theory of Didactical Situations in Mathematics*. Dordrecht: Kluwer.

Brown, J.S., Collins, A. & Duguid, S. (1989). Situated cognition and the culture of learning. In: *Educational Researcher* **18**(1), 32–42.

Burkhardt, H. & Schoenfeld, A.H. (2003). Improving educational research: Toward a more useful, more influential, and better-funded enterprise. In: *Educational Researcher* **32**(9), 3–14.

Busse, A. (2011). Upper secondary students' handling of real-world contexts. In: G. Kaiser, W. Blum, R. Borromeo Ferri & G. Stillman (Eds.), *Trends in Teaching and Learning of Mathematical Modelling (ICTMA 14)* (pp. 37–46). Dordrecht: Springer.

Carreira, S. & Baoia, A.M. (2018). Mathematical modelling with hands-on experimental tasks: On the student's sense of credibility. In: *ZDM: The International Journal on Mathematics Education* **50**(1 + 2), 201–215.

Czocher, J. (2018). How does validating activity contribute to the modeling process? In: *Educational Studies in Mathematics* **99**(2), 137–159.

DeCorte, E., Greer, B. & Verschaffel, L. (1996). Mathematics learning and teaching. In: D. Berliner & R. Calfee (Eds.), *Handbook of Educational Psychology*. New York: Macmillan.

Degrande, T., Van Hoof, J., Verschaffel, L. & Van Dooren, W. (2019). Open word problems: Taking the additive or the multiplicative road? In: *ZDM: The International Journal on Mathematics Education* **50**(1 + 2), 91–102.

Djepaxhija, B., Vos, P. & Fuglestad, A.B. (2017). Assessing mathematizing competences through multiple-choice tasks: Using students' response processes to investigate task validity. In: G. Stillman, W. Blum & G. Kaiser (Eds.), *Mathematical Modelling and Applications: Crossing and Researching Boundaries in Mathematics Education* (pp. 601–611). Cham: Springer.

Dunne, T. & Galbraith, P. (2003). Mathematical modelling as pedagogy: Impact of an immersion program. In: Q. Ye, W. Blum, S.K. Houston & Q. Jiang (Eds.), *Mathematical Modelling in Education and Culture* (pp. 16–30). Chichester: Horwood.

Francis, B. & Hobbs, D. (1991). Enterprising mathematics: A context-based course with context-based assessment. In: M. Niss, W. Blum & I. Huntley (Eds.), *Teaching of Mathematical Modelling and Applications* (pp. 147–157). Chichester: Horwood.

Frejd, P. (2013). Modes of modelling assessment: A literature review. In: *Educational Studies in Mathematics* **84**(3), 413–438.

Frejd, P. & Ärlebäck, J. (2011). First results from a study investigating Swedish upper secondary students' mathematical modelling competencies. In: G. Kaiser, W. Blum, R. Borromeo Ferri & G. Stillman (Eds.), *Trends in Teaching and Learning of Mathematical Modelling (ICTMA 14)* (pp. 407–416). Dordrecht: Springer.

Galbraith, P. & Clathworthy, N. (1990). Beyond standard models: Meeting the challenge of modelling. In: *Educational Studies in Mathematics* **21**(2), 137–163.

Galbraith, P. & Stillman, G. (2006). A framework for identifying student blockages during transitions in the modeling process. In: *ZDM: The International Journal on Mathematics Education* **38**, 143–162.

Galbraith, P., Stillman, G., Brown, J. & Edwards, I. (2007). Facilitating middle secondary modelling competencies. In: C.R. Haines, P.L. Galbraith, W. Blum & S. Khan (Eds.), *Mathematical Modelling (ICTMA 12): Education, Engineering and Economics* (pp. 130–140). Chichester: Horwood.

Gillespie, J., Binns, B., Burkhardt, H. & Swan, M. (1989). Assessment of mathematical modelling. In: W. Blum, J. Berry, R. Biehler, I.D. Huntley, G. Kaiser-Meßmer & L. Profke (Eds.), *Applications and Modelling in Learning and Teaching Mathematics* (pp. 144–152). Chichester: Horwood.

Goos, M. (2002). Understanding meta-cognitive failure. In: *Journal of Mathematical Behaviour* **21**(3), 283–302.

Greer, B. & Verschaffel, L. (2007). Modelling competencies: Overview. In: W. Blum, P.L. Galbraith, H.-W. Henn & M. Niss (Eds.), *Modelling and Applications in Mathematics Education* (pp. 219–224). New York: Springer.

Haines, C.R. (1991). Project assessment for mathematicians. In: M. Niss, W. Blum & I. Huntley (Eds.), *Teaching of Mathematical Modelling and Applications* (pp. 299–305). Chichester: Horwood.

Haines, C.R. & Crouch, R. (2007). Mathematical modelling and applications: Ability and competency frameworks. In: W. Blum, P.L. Galbraith, H-W. Henn & M. Niss (Eds.), *Modelling and Applications in Mathematics Education: The 14th ICMI Study* (pp. 417–424). New York, NY: Springer.

Haines, C.R., Crouch, R. & Davis, J. (2000). *Mathematical Modelling Skills: A Research Instrument* (Technical Report No. 55). Hatfield: University of Hertfordshire, Department of Mathematics.

Haines, C.R., Crouch, R. & Davis, J. (2001). Understanding students' modelling skills. In: J. Matos, W. Blum, K. Houston & S. Carreira (Eds.), *Modelling and Mathematics Education, ICTMA 9: Applications in Science and Technology* (pp. 366–380). Chichester: Horwood.

Haines, C.R. & Izard, J. (1995). Assessment in context for mathematical modelling. In: C. Sloyer, W. Blum & I. Huntley (Eds.), *Advances and Perspectives in the Teaching of Mathematical Modelling and Applications* (pp. 131–150). Yorklyn: Waterstreet Mathematics.

Haines, C.R., Izard, J. & Le Masurier, D. (1993). Modelling intensions realised: Assessing the full range of developed skills. In: T. Breiteig & G. Kaiser-Meßmer (Eds.), *Teaching and Learning Mathematics in Context* (pp. 200–211). Chichester: Horwood.

Hattie, J.A.C. (2009). *Visible Learning. A synthesis of over 800 Meta-Analyses Relating to Achievement*. London and New York: Routledge.

Henning, H. & Keune, M. (2007). Levels of modelling competencies. In: W. Blum, P.L. Galbraith, H-W. Henn & M. Niss (Eds.), *Modelling and Applications in Mathematics Education: The 14th ICMI Study* (pp. 225–232). New York, NY: Springer.

Hiebert, J. & Carpenter, T.P. (1992). Learning and teaching with understanding. In: D.A. Grouws (Ed.), *Handbook of Research on Mathematics Teaching and Learning* (pp. 65–97). New York: Macmillan.

Houston, S.K. (2007). Assessing the "phases" of mathematical modelling. In: W. Blum, P.L. Galbraith, H.-W. Henn & M. Niss (Eds.), *Modelling and Applications in Mathematics Education: The 14th ICMI Study* (pp. 249–256). New York, NY: Springer.

Ikeda, T. & Stephens, M. (1998). The influence of problem format on students' approaches to mathematical modeling. In: P.L. Galbraith, W. Blum, G. Booker & I. Huntley (Eds.), *Mathematical Modeling: Teaching and Assessing in a Technology-Rich World* (pp. 223–232). Chichester: Horwood.

Ikeda, T. & Stephens, M. (2001). The effects of students' discussion in mathematical modelling. In: J.F. Matos, W. Blum, S.K. Houston & S. Carreira (Eds.), *Modelling and Mathematics Education: Applications in Science and Technology* (pp. 381–390). Chichester: Horwood.

Izard, J. (2007). Assessing progress in mathematical modelling. In: C.R. Haines, P.L. Galbraith, W. Blum & S. Khan (Eds.), *Mathematical Modelling (ICTMA 12), Education, Engineering and Economics* (pp. 95–107). Chichester: Horwood.

Izard, J., Crouch, R., Haines, C.R., Houston, S.K. & Neill, N. (2003). Assessing the impact of teaching mathematical modelling: Some implications. In: S. Lamon, W.A. Parker & S.K. Houston (Eds.), *Mathematical Modelling: A Way of Life: ICTMA 11* (pp. 165–177). Chichester: Horwood.

Jankvist, U.T. & Niss, M. (2019). Upper secondary students' difficulties with mathematical modelling. In: *International Journal of Mathematical Instruction in Science and Technology*. DOI: 10.1080/0020739X.2019.1587530

Jensen, T.H. (2007). Assessing mathematical modelling competency. In: C.R. Haines, P.L. Galbraith, W. Blum & S. Khan (Eds.), *Mathematical Modelling (ICTMA 12), Education, Engineering and Economics* (pp. 141–148). Chichester: Horwood.

Julie, C. & Mudaly, V. (2007). Mathematical modelling of social issues in school mathematics in South Africa. In: W. Blum, P.L. Galbraith, H.-W. Henn & M. Niss (Eds.), *Modelling and Applications in Mathematics Education* (pp. 497–510). New York: Springer.

Kaiser, G. (2017). The teaching and learning of mathematical modeling. In: J. Cai (Ed.), *Compendium for Research in Mathematics Education* (pp. 267–291). Reston: NCTM.

Kaiser, G. & Brand, S. (2015). Modelling competencies: Past development and future perspectives. In: G. Stillman, W. Blum & M.S. Biembengut (Eds.), *Mathematical Modelling in Education Research and Practice* (pp. 129–149). Cham: Springer.

Kaiser, G. & Schwarz, B. (2010). Authentic modelling problems in mathematics: Examples and experiences. In: *Journal für Mathematik-Didaktik* 31(1), 51–76.

Kaiser-Meßmer, G. (1987). Application-oriented mathematics teaching. In: W. Blum, J.S. Berry, R. Biehler, I.D. Huntley, G. Kaiser-Messmer & L. Profke (Eds.), *Applications and Modelling in Learning and Teaching Mathematics* (pp. 66–72). Chichester: Horwood.

Kramarski, B., Mevarech, Z.R. & Arami, V. (2002). The effects of metacognitive instruction on solving mathematical authentic tasks. In: *Educational Studies in Mathematics* 49(2), 225–250.

Kunter, M. & Voss, T. (2013). The model of instructional quality in COACTIV: A multicriteria analysis. In: M. Kunter, J. Baumert, W. Blum, U. Klusmann, S. Krauss & M. Neubrand (Eds.), *Cognitive Activation in the Mathematics Classroom and Professional Competence of Teachers: Results from the COACTIV Project* (pp. 97–124). New York: Springer.

Lave, J. (1992). Word problems: A microcosm of theories of learning. In: P. Light & G. Butterworth (Eds.), *Context and Cognition: Ways of Learning and Knowing* (pp. 74–92). New York: Harvester Wheatsheaf.

Leiß, D. (2007). *Lehrerinterventionen im selbständigkeitsorientierten Prozess der Lösung einer mathematischen Modellierungsaufgabe*. Hildesheim: Franzbecker.

Leiß, D. (2010). Adaptive Lehrerinterventionen beim mathematischen Modellieren – empirische Befunde einer vergleichenden Labor-und Unterrichtsstudie. In: *Journal für Mathematik-Didaktik* 31(2), 197–226.

Leikin, R. & Levav-Waynberg, A. (2007). Exploring mathematics teacher knowledge to explain the gap between theory-based recommendations and school practice in the use of connecting tasks. In: *Educational Studies in Mathematics* 66, 349–371.

Leiss, D., Schukajlow, S., Blum, W., Messner, R. & Pekrun, R. (2010). The role of the situation model in mathematical modelling: Task analyses, student competencies, and teacher interventions. In: *Journal für Mathematik-Didaktik* 31(1), 119–141.

Lesh, R. & Doerr, H. (2003). *Beyond Constructivism: Models and Modeling Perspectives on Mathematics Teaching, Learning and Problem Solving*. Mahwah, NJ: Lawrence Erlbaum.

Lesh, R. & Harel, G. (2003). Problem solving, modeling, and local conceptual development. In: *Mathematical Thinking and Learning* 5(23), 157–189.

Lesh, R., Hoover, M., Hole, B., Kelly, A. & Post, T. (2000). Principles for developing thought-revealing activities for students and teachers. In: A. Kelly & R. Lesh (Eds.), *Handbook of Research Design in Mathematics and Science Education* (pp. 541–646). Mahwah, NJ: Lawrence Erlbaum.

Lesh, R. & Zawojewski, J. (2007). Problem solving and modeling. In: F. Lester (Ed.), *Second Handbook of Research on Mathematics Teaching and Learning* (pp. 763–802). Greenwich: Information Age Publishing.

Ludwig, M. & Xu, B. (2010). A comparative study of modelling competencies among Chinese and German students. In: *Journal für Mathematik-Didaktik* 31(1), 77–97.

Maaß, K. (2007). Modelling in class: What do we want the students to learn? In: C.R. Haines, P.L. Galbraith, W. Blum & S. Khan (Eds.), *Mathematical Modelling: Education, Engineering and Economics* (pp. 63–78). Chichester: Horwood.

Maaß, K. & Engeln, K. (2018). Impact of professional development involving modelling on teachers and their teaching. In: *ZDM: The International Journal on Mathematics Education* 50(1 + 2), 273–285.

Matos, J.F. & Carreira, S. (1997). The quest for meaning in students' mathematical modelling activity. In: S.K. Houston, W. Blum, I.D. Huntley & N.T. Neill (Eds.), *Teaching & Learning Mathematical Modelling* (pp. 63–75). Chichester: Horwood.

Money, R. & Stephens, M. (1993). Linking applications modelling and assessment. In: J. de Lange, I. Huntley, C. Keitel & M. Niss (Eds.), *Innovation in Maths Education by Modelling and Applications* (pp. 323–336). Chichester: Horwood.

Naylor, T. (1991). Assessment of a modelling and applications teaching module. In: M. Niss, W. Blum & I. Huntley (Eds.), *Teaching of Mathematical Modeling and Applications* (pp. 317–325). Chichester: Horwood.

Nesher, P. (1980). The stereotyped nature of school word problems. In: *For the Learning of Mathematics* **1**(1), 41–48.

Newell, A. & Simon, H.A. (1971). *Human Problem Solving*. Englewood Cliffs, NJ: Prentice-Hall.

Niss, M. (1993). Assessment of mathematical applications and modelling in mathematics teaching. In: J. de Lange, I. Huntley, C. Keitel & M. Niss (Eds.), *Innovation in Maths Education by Modelling and Applications* (pp. 41–51). Chichester: Horwood.

Niss, M. (1999). Aspects of the nature and state of research in mathematics education. In: *Educational Studies in Mathematics* **40**(1), 1–24.

Niss, M. (2010). Modeling a crucial aspect of students' mathematical modeling. In: R. Lesh, P.L. Galbraith, C.R. Haines & A. Hurford (Eds.), *Modeling Students' Mathematical Modeling Competencies: ICTMA 13* (pp. 43–59). New York: Springer.

Niss, M. (2017). Obstacles related to structuring for nathematization encountered by students when solving physics problems. In: *International Journal of Science and Mathematics Education* **15**, 1441–1462.

Niss, M. & Jensen, T.H. (2002). *Kompetencer og matematiklæring – Ideer og inspiration til udvikling af matematikundervisning i Danmark*, number 18 in Uddannelsesstyrelsens temahæfteserie. Copenhagen: The Ministry of Education.

Niss, M. & Højgaard, T. (2011). *Competencies and Mathematical Learning*. Roskilde: Roskilde University.

OECD (2013). *PISA 2012 Results: What Students Know and Can Do* (Vol. 1). Paris: OECD Publishing.

OECD (2016). *PISA 2015 Results: Excellence and Equity in Education* (Vol. 1). Paris: OECD Publishing.

OECD (2019). *PISA 2018 Results (Volume I): What Students Know and Can Do*. Paris: OECD Publishing. https://doi.org/10.1787/5f07c754-en

Ottesen, J. (2001). Do not ask what mathematics can do for modelling. In: D. Holton (Ed.), *The Teaching and Learning of Mathematics at University Level* (pp. 335–346). Dordrecht: Kluwer.

Palm, T. (2007). Features and impact of the authenticity of applied mathematical school tasks. In: W. Blum, P.L. Galbraith, H.-W. Henn & M. Niss (Eds.), *Modelling and Applications in Mathematics Education* (pp. 201–208). New York: Springer.

Plath, J. & Leiß, D. (2018). The impact of linguistic complexity on the solution of mathematical modelling tasks. In: *ZDM: The International Journal on Mathematics Education* **50**(1 + 2), 159–171.

Polya, G. (1973). *How to Solve It: A New Aspect of Mathematical Methods*. Princeton, NJ: Princeton University Press.

Potari, D. (1993). Mathematisation in a real-life investigation. In: J. de Lange, C. Keitel, I. Huntley & M. Niss (Eds.), *Innovation in Mathematics Education by Modelling and Applications* (pp. 235–243). Chichester: Horwood.

Rellensmann, J., Schukajlow, S. & Leopold, C. (2017). Make a drawing: Effects of strategic knowledge, drawing accuracy, and type of drawing on students' mathematical modelling performance. In: *Educational Studies in Mathematics* **95**, 53–78.

Reusser, K. (2001). Co-constructivism in educational theory and practice. In: N.J. Smelser, P. Baltes & F.E. Weinert (Eds.), *International Encyclopedia of the Social and Behavioral Sciences* (pp. 2058–2062). Oxford: Pergamon and Elsevier Science.

Reusser, R. & Stebler, R. (1997). Every word problem has a solution: The suspension of reality and sense-making in the culture of school mathematics. In: *Learning and Instruction* **7**, 309–328.

Rittle-Johnson, B. & Star, J.R. (2009). Compared with what? The effects of different comparisons on conceptual knowledge and procedural flexibility for equation solving. In: *Journal of Educational Psychology* **101**(3), 529–544.

Schaap, S., Vos, P. & Goedhart, M. (2011). Students overcoming blockages while building a mathematical model: Exploring a framework. In: G. Kaiser, W. Blum, R. Borromeo Ferri & G. Stillman (Eds.), *Trends in Teaching and Learning of Mathematical Modelling* (pp. 137–146). New York: Springer.

Schoenfeld, A.H. (1985). *Mathematical Problem Solving*. New York: Academic Press.

Schoenfeld, A.H. (1987). What's all the fuss about metacognition? In: A.H. Schoenfeld (Ed.), *Cognitive Science and Mathematics Education* (pp. 189–215). Hillsdale, NJ: Lawrence Erlbaum Associates.

Schoenfeld, A.H. (1991). On mathematics as sense-making: An informal attack on the unfortunate divorce of formal and informal mathematics. In: J.F. Voss, D.N. Perkins & J.W. Segal (Eds.), *Informal Reasoning and Education* (pp. 311–343). Hillsdale: Erlbaum.

Schoenfeld, A.H. (1994). *Mathematical Thinking and Problem Solving*. Hillsdale: Erlbaum.

Schukajlow, S. (2011). *Mathematisches Modellieren. Schwierigkeiten und Strategien von Lernenden als Bausteine einer lernprozessorientierten Didaktik der neuen Aufgabenkultur*. Münster: Waxmann.

Schukajlow, S., Kaiser, G. & Stillman, G. (2018). Empirical research on teaching and learning of mathematical modelling: A survey of the state-of-the-art. In: *ZDM: The International Journal on Mathematics Education* **50**(1 + 2), 5–18.

Schukajlow, S. & Krug, A. (2013). Planning, monitoring and multiple solutions while solving modelling problems. In: A. Lindmeier & A. Heinze (Eds.), *Mathematics Learning across the Life Span: Proceedings of the 37th Conference of the International Group for the Psychology of Mathematics Education* (pp. 177–184). Kiel: IPN.

Schukajlow, S., Krug, A. & Rakoczy, K. (2015). Effects of prompting multiple solutions for modelling problems on students' performance. In: *Educational Studies in Mathematics* **89**(3), 393–417.

Schukajlow, S., Leiss, D., Pekrun, R., Blum, W., Müller, M. & Messner, R. (2012). Teaching methods for modelling problems and students' task-specific enjoyment, value, interest and self-efficacy expectations. In: *Educational Studies in Mathematics* **79**(2), 215–237.

Sol, M., Giménez, J. & Rosich, N. (2011). Project modelling routes in 12–16-year-old pupils. In: G. Kaiser, W. Blum, R. Borromeo Ferri & G. Stillman (Eds.), *Trends in Teaching and Learning of Mathematical Modelling (ICTMA 14)* (pp. 231–240). Dordrecht: Springer.

Stender, P. (2018). The use of heuristic strategies in modelling activities. In: *ZDM: The International Journal on Mathematics Education* **50**(1 + 2), 315–326.

Stender, P. & Kaiser, G. (2016). Fostering modeling competencies for complex situations. In: C. Hirsch & A. Roth McDuffie (Eds.), *Mathematical Modeling and Modeling Mathematics* (pp. 107–115). Reston: NCTM.

Stender, P., Krosanke, N. & Kaiser, G. (2017). Scaffolding complex modelling problems: An in-depth Study. In: G.A. Stillman, W. Blum & G. Kaiser (Eds.), *Mathematical Modelling and Applications: Crossing and Researching Boundaries in Mathematics Education* (pp. 467–477). Cham: Springer.

Stillman, G. (1998). Engagement with task context of applications task: Student performance and teacher beliefs. In: *NOMAD* **6**(3/4), 51–70.
Stillman, G. (2000). Impact of prior knowledge of task context on approaches to application tasks. In: *Journal of Mathematical Behaviour* **19**(31), 333–361.
Stillman, G. (2011). Applying metacognitive knowledge and strategies in applications and modelling tasks at secondary school. In: G. Kaiser, W. Blum, R. Borromeo Ferri & G. Stillman (Eds.), *Trends in Teaching and Learning of Mathematical Modelling (ICTMA 14)* (pp. 165–180). Dordrecht: Springer.
Stillman, G. (2015). Applications and modeling research in secondary classrooms: What have we learnt? In: S. J. Cho (Ed.), *Selected Regular Lectures from the 12th International Congress on Mathematics Education* (pp. 791–806). Cham: Springer.
Stillman, G. (2019). State of the art on modelling in mathematics education: Lines of inquiry. In: G. Stillman & J. Brown (Eds.), *Lines of Inquiry of Mathematical Modelling Research in Education* (pp. 1–19). Cham: Springer.
Stillman, G. & Brown, J. (2014). Evidence of implemented anticipation in mathematising by beginning modelers. In: *Mathematics Education Research Journal* **26**, 763–789.
Stillman, G., Brown, J. & Galbraith, P. (2010). Identifying challenges within transition phases of mathematical modeling activities at year 9. In: R. Lesh, P.L. Galbraith, C.R. Haines & A. Hurford (Eds), *Modeling Students' Mathematical Modeling Competencies* (pp. 385–398). New York: Springer.
Stillman, G., Brown, J. & Geiger, V. (2015). Facilitating mathematisation in modelling by beginning modellers in secondary school. In: G. Stillman, W. Blum & M.S. Biembengut (Eds.), *Mathematical Modelling in Education Research and Practice* (pp. 93–104). Cham: Springer.
Stillman, G. & Galbraith, P. (1998). Applying mathematics with real world connections: Metacognitive characteristic of secondary students. In: *Educational Studies in Mathematics* **36**(2), 157–195.
Stohlmann, M., DeVaul, L., Allen, C., Adkins, A., Ito, T., Lockett, D. & Wong, N. (2016). What is known about secondary grades mathematical modelling: A review. In: *Journal of Mathematics Research* **8**(5), 12–28.
Tanner, H. & Jones, S. (1995). Developing metacognitive skills in mathematical modelling: A socio-constructivist interpretation. In: C. Sloyer, W. Blum & I. Huntley (Eds.), *Advances and Perspectives in the Teaching of Mathematical Modelling and Applications* (pp. 61–70). Yorklyn: Water Streets Mathematics.
Timperley, H.S. (2011). *Realizing the Power of Professional Learning*. London: Open University Press.
Tsamir, P., Tirosh, D., Tabach, M. & Levenson, E. (2010). Multiple solution methods and multiple outcomes: Is it a task for kindergarten children? In: *Educational Studies in Mathematics* **73**, 217–231.
Turner, R. (2007). Modelling and applications in PISA. In: W. Blum, P.L. Galbraith, H.-W. Henn & M. Niss (Eds.), *Modelling and Applications in Mathematics Education: The 14th ICMI Study* (pp. 433–440). New York, NY: Springer.
Verschaffel, L., Greer, B. & DeCorte, E. (2000). *Making Sense of Word Problems*. Lisse: Swets&Zeitlinger.
Verschaffel, L., Van Dooren, W., Greer, B. & Mukhopadhyay, S. (2010). Reconceptualising word problems as exercises in mathematical modeling. In: *Journal für Mathematik-Didaktik* **31**(1), 9–29.
Vom Hofe, R. & Blum, W. (2016). "Grundvorstellungen" as a category of subject-matter didactics. In: *Journal für Mathematik-Didaktik* **37**(Supplement 1), 225–254.

Vorhölter, K. (2018). Conceptualization and measuring of metacognitive modelling competencies: Empirical verification of theoretical assumptions. In: *ZDM: The International Journal on Mathematics Education* **50**(1 + 2), 343–354.

Vorhölter, K., Krüger, A. & Wendt, L. (2019). Metacognition in mathematical modeling: An overview. In: S.A. Chamberlin & B. Sriraman (Eds.), *Affect in Mathematical Modeling* (pp. 29–51). Cham: Springer.

Vos, P. (2007). Assessment of applied mathematics and modelling: Using a laboratory-like environment. In: W. Blum, P.L. Galbraith, H.-W. Henn & M. Niss (Eds.), *Modelling and Applications in Mathematics Education: The 14th ICMI Study* (pp. 441–448). New York: Springer.

Watson, A. (Ed.). (1998). *Situated Cognition and the Learning of Mathematics.* Oxford: University of Oxford.

Werle de Almeida, L.M. (2018). Considerations on the use of mathematics in modelling activities. In: *ZDM: The International Journal on Mathematics Education* **50**(1 + 2), 19–30.

Xin, Z., Lin, C., Zhang, L. & Yan, R. (2007). The performance of Chinese primary school students on realistic arithmetic word problems. In: *Educational Psychology in Practice* **23**, 145–159.

Zöttl, L. Ufer, S. & Reiss, K. (2011). Assessing modelling competencies using a multidimensional IRT-approach. In: G. Kaiser, W. Blum, R. Borromeo Ferri & G. Stillman (Eds.), *Trends in Teaching and Learning of Mathematical Modelling* (pp. 427–437). New York, NY: Springer.

7
CASES OF MATHEMATICAL MODELLING FROM EDUCATIONAL PRACTICES

7.1 Introduction

This chapter presents and discusses selected successful cases of the teaching and learning of mathematical modelling from actual educational practices at different educational levels from around the world (Australia, Denmark, England, Germany, USA). By "successful" we mean that expectations held by those who implemented the cases in practice have, according to their reports, been fulfilled to a satisfactory degree, meaning students have been engaged in genuine modelling activities and have constructed appropriate models; progression in students' modelling competency/ies have been detected as a result of their involvement in modelling programmes over a longer period of time; or teachers have reported on repeatedly fruitful implementation of certain materials in everyday classrooms.

We begin by taking a closer look at what counts as a *case*. We use the term "case" to denote an individual example – a constellation of objects, a set of circumstances, a situation or a process – which is an example of something, i.e., a special instance of a larger class of entities that share some characteristic features or properties. In other words, a case is not just a singular stand-alone example; rather, it points to and is a representative of something bigger than itself. In this chapter, we shall consider three types of cases concerning modelling in mathematics education. The first type of cases consists of individual modelling examples that have been implemented in actual teaching and learning and have been reported and investigated or evaluated with respect to the context(s) in which the implementation took place. The second type consists of sustained curricular programmes of mathematical modelling put into practice in certain educational settings, e.g., institutions or countries, during some extended period either within one institution or across different institutions. The third type of case consists of educational materials that have been designed for and implemented in a variety of different settings.

146 Cases of mathematical modelling

We have provided a few cases of every type, for a total seven cases. We made no attempt to offer an exhaustive selection of significant cases – not even a balanced one. Rather, our goal is to point to the existence of different types of cases of modelling that have been implemented in the actual teaching and learning of mathematics.

7.2 Individual implemented modelling examples

In this section, we present two modelling examples that have been treated by or with secondary students within the framework of projects to gain empirical insight into possibilities and conditions for the implementation of modelling in everyday mathematics classrooms. These examples are described in considerable detail.

Case 1: Cable drum

The problem

A modelling example that has been treated both by school students (grade 9 onwards) and in teacher education is shown in Figure 7.1 (original German version in Förster & Herget, 2002).

The question is: "How long is the cable that fits on the drum shown on the photo, and whose data are given?". In the following, we draw on considerations and experiences reported in Förster and Kaiser (2010) and Förster (2018).

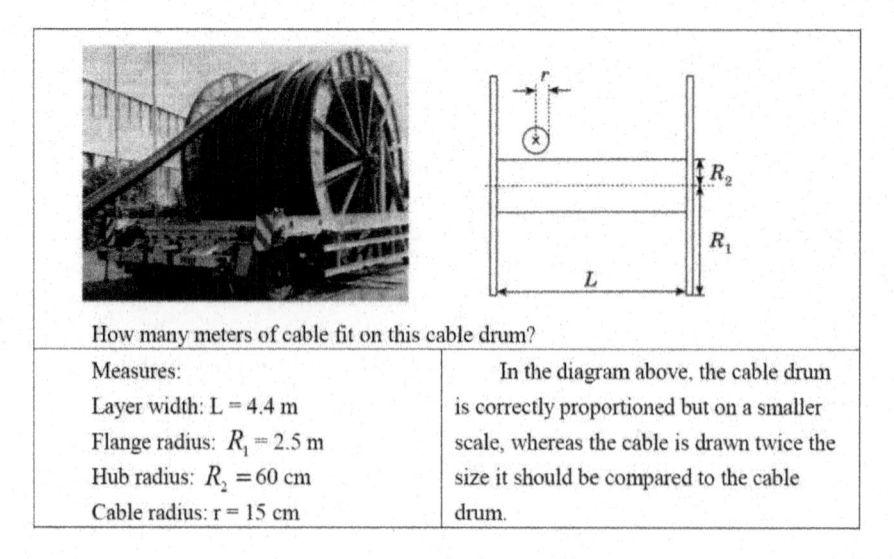

How many meters of cable fit on this cable drum?	
Measures: Layer width: L = 4.4 m Flange radius: R_1 = 2.5 m Hub radius: R_2 = 60 cm Cable radius: r = 15 cm	In the diagram above, the cable drum is correctly proportioned but on a smaller scale, whereas the cable is drawn twice the size it should be compared to the cable drum.

FIGURE 7.1 The working sheet "cable drum"

Source: Förster & Kaiser, 2010

Typically, students or teachers working on this task produce many different results for the cable length, stemming from different models of the ways in which the cable is being wound and its length calculated but also from certain mistakes made during the modelling process. Förster reports that in a German grade 10 class, the students ended up with 20 different lengths, ranging from 442 m to 1266 m.

The following five types of models are particularly natural to come up with. When we report percentages, we refer to the figures given in Förster (2018), based on observations of 449 ninth and tenth graders (15- to 16-year-old students). About a quarter of these students found no approach at all without support from the teacher.

Model M1: Separated circles

This is the most popular model with students as well as with teachers (about 30% of the students observed chose this model). It consists of a description of the cross-section of the cable wound on the drum using several layers of circles (see Figure 7.2; all of the following drawings are taken from ninth and tenth grade classrooms). It is presupposed here (and in all other models as well) that the cable is flexible enough to be wound around the drum.

In this model, the cable is perceived as consisting of many separated circular pieces of cable (annuli), with the same number of pieces in each layer. How many such pieces fit onto the drum? See Figure 7.3: Since $L:2r \approx 14.7$, 14 circles fit horizontally, and since $(R_1 - R_2):2r \approx 6.3$, 6 layers fit vertically. Therefore, the cable

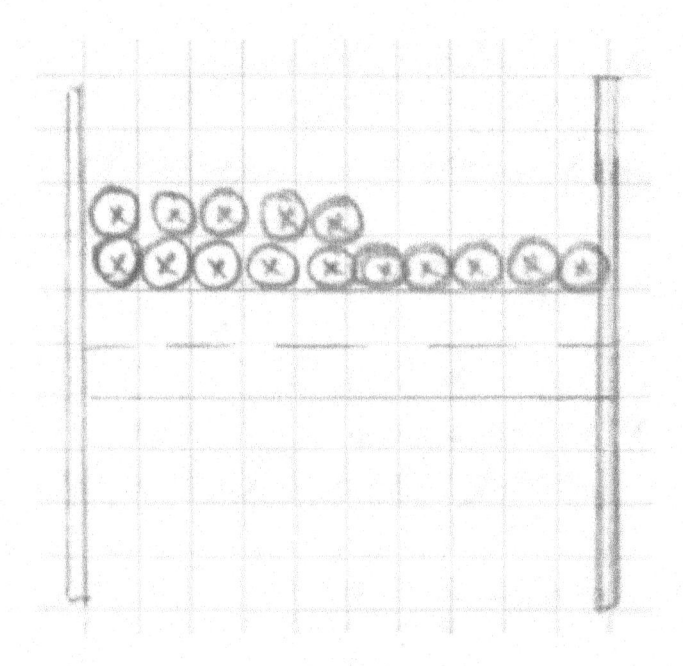

FIGURE 7.2 The circles model M1

148 Cases of mathematical modelling

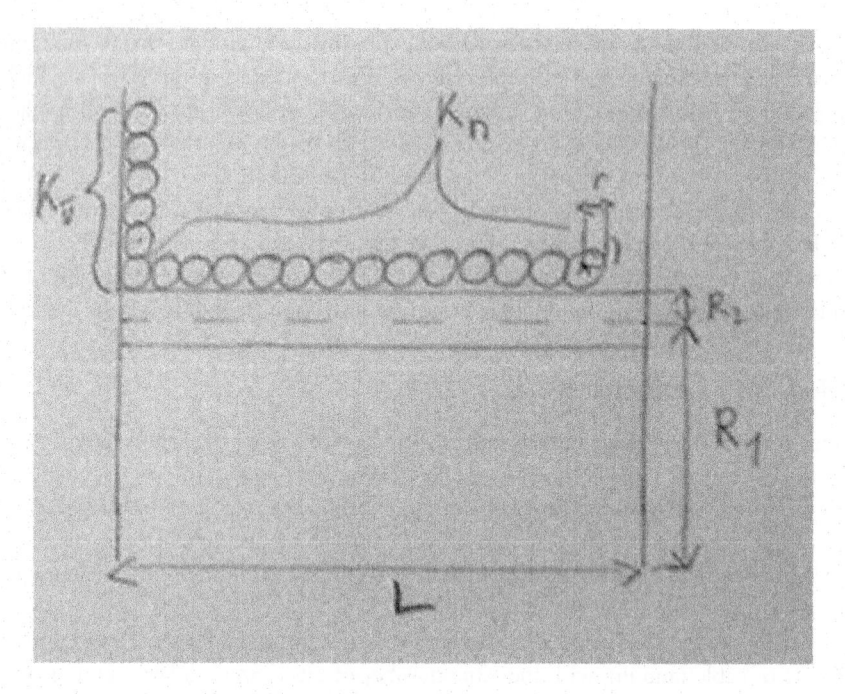

FIGURE 7.3 The circles model with given quantities

consists of 84 separated circular pieces, of which exactly the 14 pieces in one layer have the same radius. If the cable were allowed to protrude a bit above the drum, then seven layers may fit onto it. However, this would most probably not be realised, both for security reasons and because the full cable drum could then not be rolled on the ground.

The lengths of the various circular pieces can be calculated; see Table 7.1 (rounded off to centimetres). The common radius of the 14 smallest annuli is $R_2 + r = 75$ cm, the radius of the 14 annuli in the second layer is $R_2 + 3r = 105$ cm, etc., until the sixth layer with radius $R_2 + 11r = 225$ cm, if we measure the radii of these annuli in the *centre* of the cable (*model M1a*). The length of one annulus in the first layer is $2\pi \cdot 75$ cm ≈ 471 cm, etc. With six layers, the final result for the cable length is approximately 792 m. If a seventh layer is considered, the cable length is approximately 1016 m.[1]

If the radii of the circular pieces are not measured in the centre but at the lower "skin side" of the cables (*inner* radii, *model M1b*), these are, in all layers, 15 cm less than in Table 7.1 (this type of calculation was used by 10% of the students observed). This certainly leads to an under-estimate of the real length of the cable. Table 7.2 shows the calculations analogous to Table 7.1. With 6 layers, the final result for the cable length is now approximately 713 m, with seven layers approximately 924 m.

TABLE 7.1 Lengths of the single layers for the circles model (all measures in m)

Layer	Radius	Circle length	Length of layer	Accumulated cable length
1	0.75	4.71	65.97	65.97
2	1.05	6.60	92.36	158.34
3	1.35	8.48	118.75	277.09
4	1.65	10.37	145.14	422.23
5	1.95	12.25	171.53	593.76
6	2.25	14.14	197.92	**791.68**
7	2.55	16.02	224.31	1015.99

TABLE 7.2 Lengths of the layers with inner radii (all measures in m)

Layer	Radius	Circle length	Length of layer	Accumulated cable length
1	0.6	3.77	52.78	52.78
2	0.9	5.65	79.17	131.95
3	1.2	7.54	105.56	237.50
4	1.5	9.42	131.95	369.45
5	1.8	11.31	158.33	527.79
6	2.1	13.19	184.73	**712.51**
7	2.4	15.08	211.12	923.63

TABLE 7.3 Lengths of the layers with outer radii (all measures in m)

Layer	Radius	Circle length	Length of layer	Accumulated cable length
1	0.9	5.65	79.17	79.17
2	1.2	7.54	105.56	184.73
3	1.5	9.42	131.95	316.67
4	1.8	11.31	158.34	475.01
5	2.1	13.19	184.73	659.73
6	2.4	15.08	211.11	**870.85**
7	2.7	16.96	237.50	1108.35

If the radii are measured at the upper "skin side" of the cables (*outer* radii, *model M1c*), these are, in all layers, 15 cm larger than when measured in the centre (this type of calculation was used by 5% of the students observed). Table 7.3 shows the calculations for this version of the model. Now, the length of the cable is approximately 871 m with six layers and approximately 1108 m with seven layers.

Model M2: Helices

Some modellers find a (seemingly) better model where the cable is, in each layer, wound in the shape of a helix (see Figure 7.4 where the first layer is drawn).

According to this model, the cable becomes a bit longer since each circular piece according to model M1 must be mentally transformed into a part of a helix. However, calculations using the Pythagorean theorem show that the difference to the first circles model is minimal. Let us calculate, as an example, the length of one helix winding in the first layer. For this purpose, we must unroll the cable; see Figure 7.5.

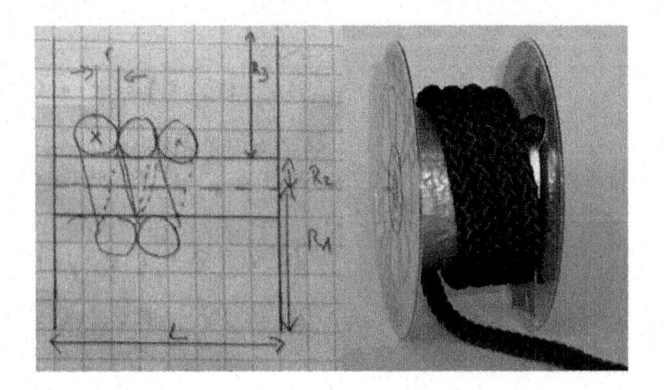

FIGURE 7.4 The helix model M2
Source: Photo by Frank Förster

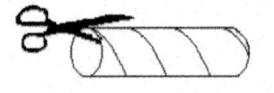

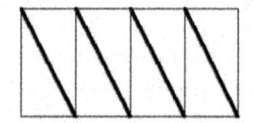

FIGURE 7.5 Unrolling the helix

The length of the first helix is:

$$\sqrt{(2\pi \cdot 75\,\text{cm})^2 + (30\,\text{cm})^2} \approx 472 \text{ cm},$$

i.e., the helix is only 1 cm longer than the circle in model M1. The differences become slightly bigger from layer to layer, but they are still rather negligible. With 6 layers the resulting length of the cable is, rounded off, again approximately 792 m. Therefore, this technical complication is, in retrospect, not worth the effort.

Model M3: shifted separated circles "on gaps"

Practising and prospective teachers tend to use, often as a subsidiary improved model, the idea of "winding the cable on gaps", so the circular cable pieces are shifted from layer to layer; see Figure 7.6.

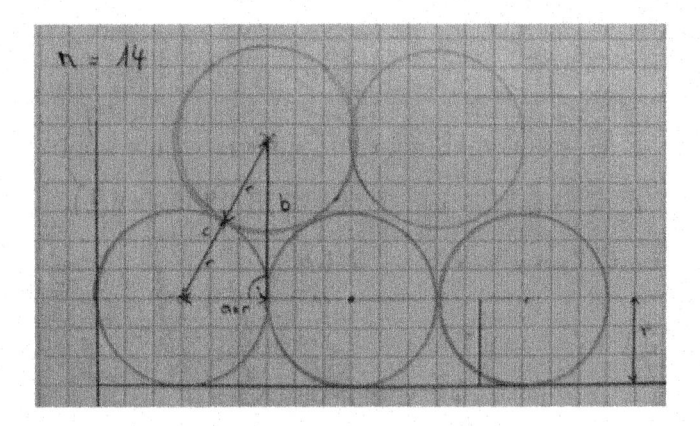

FIGURE 7.6 The shifted circles "on gaps" model M3

The reason to choose this model is that the vertical distance between two cable centres is now smaller than $2r = 30$ cm, which was the distance in the previous models. By means of the Pythagorean theorem, we calculate the new vertical distance between two cable centres to:

$\sqrt{(2r)^2 - r^2} = \sqrt{3} \cdot r = \sqrt{3} \cdot 15 \, \text{cm} \approx 26$ cm.

Since the layers are nearer to each other, 7 layers will now fit on the drum because $(R_1 - R_2) : \sqrt{3}r \approx 7.3$. Thus, from the second layer on, the circular pieces wound around the hub of the drum will become a bit smaller; in the end, the seventh layer gives rise to an advantage. Table 7.4 shows the new calculations.

The result for the length of the cable is now approximately **942 m**. This model is sometimes used qualitatively rather than quantitatively, with the following argument: When using this kind of winding, seven layers will indeed fit on the drum. Compared to the "pure" circles model, the windings become shorter, so 1016 m

TABLE 7.4 Lengths of the single layers in the shifted circles "on gaps" model (all measures in m)

Layer	Radius	Circle length	Length of layer	Accumulated cable length
1	0.7500	4.71	65.97	65.97
2	1.0098	6.34	88.83	154.80
3	1.2696	7.98	111.68	266.48
4	1.5294	9.61	134.54	401.02
5	1.7892	11.24	157.39	558.41
6	2.0490	12.87	180.24	738.65
7	2.3088	14.51	203.10	**941.75**

152 Cases of mathematical modelling

(the length in the circles model M1a for 7 layers) will certainly be too much, but 792 m (the length in model M1a for 6 layers) will certainly be too little, so a reasonable estimate will be somewhere in the middle, around 900 m.

However, this "improved" model does not quite fit the real situation because cables cannot be wound like that (the reader may try it by him/herself with an easily accessible cable, for example, from a lamp – unplug before trying!); see Figure 7.7. In real windings, each layer consists of a helix, and the helices have different directions from layer to layer, so the cable cannot be laid "on gaps"; the helix in the next layer lies entirely above the previous helix, like in the seemingly simpler circles model.

FIGURE 7.7 A real wound cable
Source: Photo by Frank Förster

Therefore, model 3, though mathematically more elaborate, is, unfortunately, not suitable.

With this insight in mind, we come back to our first model. The most suitable model in terms of real cables is a variant of model 1 (*model M1**) with layers of circles where each layer is shifted against the previous layer, as in model M3, but the distance between neighbouring layers is 30 cm, as in models M1 and M2, because, as explained above, each cable in the next layer has to be laid above the cables beneath it (see Figure 7.8).

Since each layer consists of 14 circles, this model variant M1* leads to exactly the same result as model M1. Referring to model M1a (radii measured in the centres), the result of model M1* is approximately 792 m or, rounded off more liberally, approximately 800 m.

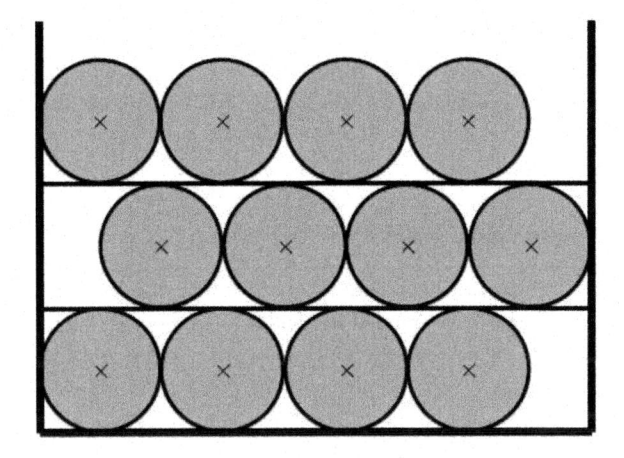

FIGURE 7.8 The shifted circles model M1★

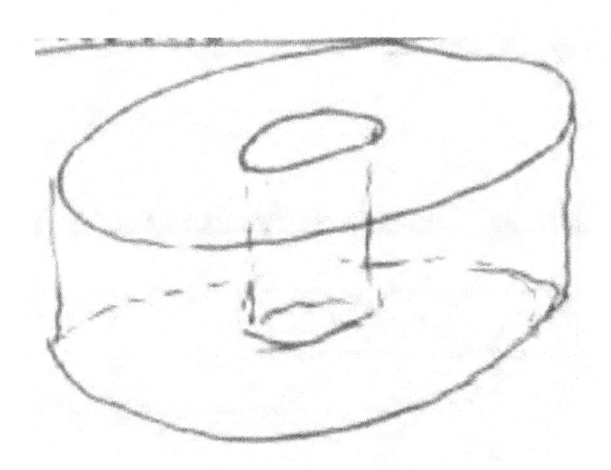

FIGURE 7.9 The volume model M4

Model M4: volume without holes

Instead of modelling the actual winding of the cable on the drum, the following model uses the volume of the cable (see Figure 7.9).

We calculate the cable length in this model by dividing the volume of the cable by its cross section. The simplest volume model (used by more than 10% of all students observed) presupposes that the cable fills the space densely on the drum, which means that there is no cable-free space (no "holes"). Its total volume is:

$(\pi \cdot R_1^2 - \pi \cdot R_2^2) \cdot L \approx 81.42$ m^3.

154 Cases of mathematical modelling

The cross section of the cable is $\pi \cdot r^2 \approx 0.0707$ m^2, thus the cable length is approximately 1152 m. It is clear from the construction (cable as a solid without holes) that this result is an upper bound for the real length. Many students and teachers who solved this problem used this as a second model after the circles model M1 to get such an upper bound.

Model M5: volume with holes

Considering the cable-free space (the "holes") in the solid caused by the cylindrical cable according to the circles model M1 (see Figure 7.2), the volume is multiplicatively reduced by the ratio of the area of a circle and the area of a surrounding square, that is, by the factor $\pi/4 \approx 0.7854$ (this approach was used by more than 10% of all students observed). This leads to a cable length of approximately $0.7854 \cdot 1152$ m ≈ 905 m.

Considering the holes according to the shifted circles model M3, the reduction factor is now the ratio of the area of a semicircle and the area of a suitable equilateral triangle, that is, $\pi/(2 \cdot \sqrt{3}) \approx 0.9069$, for the four middle layers, whereas the ratio is $\pi/4 \approx 0.7854$ as before for the two outer layers (see Figure 7.10). The average reduction factor is 0.8259.

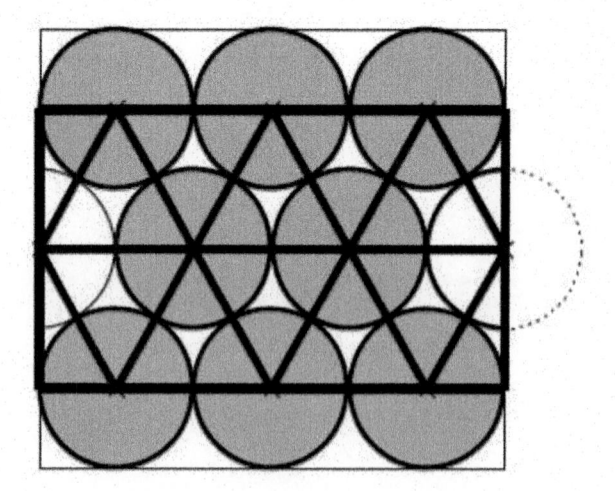

FIGURE 7.10 The gaps in the volume model

The resulting new cable length is approximately $0.8259 \cdot 1152$ m ≈ 951 m.

Evaluating the models

A natural validation of the outcomes of the various models would be to confront them with reality: How long is the actual cable on this drum? In some cases, this will be possible by simply reading what is written on the drum, but if we have no other information than the measures given on the working sheet in Figure 7.1, we

Cases of mathematical modelling **155**

are forced to validate the model results by plausible reasoning: How far away from real cable windings are the different ways in which the cable is wound on the drum? Which results are certainly too small (such as the 712 m from model M1b because the cable is longer than the length of its inner surface)? Which results are certainly too large (such as the 1152 m from model M4 because there are substantial cable-free "holes" and also the 1108 m from model 1c with 7 layers and measures based on the length of the outer cable surface)? Therefore, any result in a range between 750 m and 1050 m seems plausible.

The "cable drum" in the classroom

Treating the "cable drum" problem in the classroom may stimulate many activities. The situation is easily comprehensible by students, many of whom may already have seen a real cable drum on a construction site (like in Figure 7.11), for instance, if telephone cables are being put in the ground.

That is why the search for an appropriate geometric model can start immediately after reading the task. It is advisable to let students first estimate on a "gut feeling" level how long the cable might be. The solution process allows for natural internal differentiation within the classroom insofar as each student, or a group of students, can build a model corresponding to his/her mathematical abilities. It is also possible to find two or more different ways to model the problem, where the second way

FIGURE 7.11 A cable drum in roadwork

156 Cases of mathematical modelling

may serve as validation of the first solution, for instance, if the volume model M4 is used to discuss whether the result according to the circles model M1 is plausible. Since circles play a decisive role, this topic should be known by the students. The task is usually suitable from grade 9 onwards, depending on the treatment of circles in the curriculum. Apart from fostering modelling competencies and demonstrating that mathematics is present in the real world, the "cable drum" task can also serve as an environment for rehearsing and practising the topic area of circles (basic notions, circumference, area) and cylinders.

Experiences in the classroom show that the "cable drum" task is motivating for students, although it is most certainly not relevant for students' current everyday or future professional lives and it is not authentic since in the real world it will generally be known how long the cable is (perhaps it is even written on the drum). However, the task is both challenging and manageable, and the modelling activities that take place, individually or in groups, are interesting and motivating *per se*, in addition to giving access to the kinds of – authentic – considerations that occur in real modelling. Experiences in many classrooms show that working on this task in groups leads to particularly stimulating discussions about reasonable approaches. By confronting the students' models with reality (does a real cable actually consist of singular annuli or helices?) they learn essential features of modelling in a natural way by dealing with this task.

The discussion of the shifted circles model M3 is particularly instructive. Students may believe that they have found a better model than the one merely based on layers of circles, and perhaps they even think they have found "the optimal" model. Below is a typical student argument for switching from M1 to M3 (Figure 7.12).

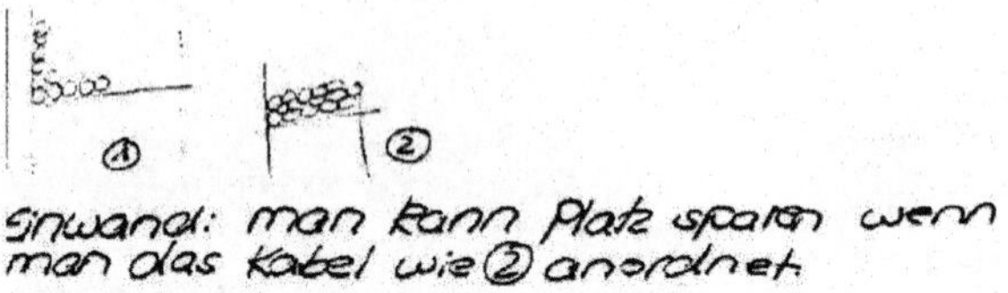

FIGURE 7.12 A seeming model improvement

("Objection: You can save space if you arrange the cable like in ②"). Classroom experiences show that students who have developed this model are generally so enthusiastic and proud about the seeming model improvement that they do not validate their results. The teacher (respectively teacher educator, if the learners are pre- or in-service teachers) should stimulate validation and evaluation activities, either by thought experiments (for strong visual minds) or – even better – by real experiments with cables, wires or ropes, to let the students discover that model M3 is inappropriate. Fortunately, in a rich teaching-learning environment, students are not, in general, frustrated when they realise that their carefully designed model does not work because they acknowledge that they have gained some valuable insight instead. This example shows that a mathematical model must be evaluated not according to its beauty or mathematical sophistication but according to its usefulness in dealing with the real world. This is an important general insight about mathematical modelling, which is part of meta-knowledge of mathematics as a discipline and which the cable drum example generates almost automatically.

During the modelling process, many instructive mistakes can occur in the classroom. In a case study with 200 lower secondary students (15- to 16-year-olds), Förster (2018) found many different mistakes; here is a selection:

- A sequence of winding radii in model M1a with a distance of $r = 15$ cm instead of 30 cm, beginning with $R_2 + r/2 = 67.5$ cm instead of $R_2 + r = 75$ cm, so 67.5 cm, 82.5 cm, etc., resulting in a cable length of 554 m;
- A sequence of radii in model M1a with a distance of $r = 15$cm, correctly beginning with $R_2 + r = 75$ cm, so first 75 cm but then 90 cm, etc., resulting in a cable length of 583 m with 6 layers and 692 m with 7 layers;
- The same wrong model as the previous one, but now with 8 layers because of an erroneous calculation of the number of layers by $R1:2r \approx 8.3$, resulting in a cable length of 791 m;
- A sequence of winding radii in model M1b with a distance of $4r = 60$ cm, correctly beginning with 75 cm and using 6 layers, resulting in a cable length of 1108 m;
- Calculating the circumferences of the circles in the circles model by using the sum of the inner and the outer radii, resulting in a cable length for 6 layers of 1582 m.

In addition, numerous differing final results occurred because of inappropriate rounding of numbers. An observation in nearly all classrooms was that students are obviously not used to rounding off for the final result. Figure 7.13 shows a typical solution (using model M1a) where the result is given with an accuracy of three decimal points, which means precision to the millimetre. Apart from that, the solution is correct and clearly presented.

The "cable drum" problem in its present form appears to be rather "closed" because all relevant data are given. A more open version where some data are not given but must be inferred from the picture may seem worthwhile. Another possibility is to bring a cable on a drum into the classroom, for instance, a garden hose drum, and ask the same question as before. Here the students must identify and afterwards measure the decisive quantities. However, the experiences reported in Förster (2018) indicate that the present version of the task is still open enough to stimulate modelling activities and that more open versions may sometimes lead to arbitrary approaches with no results, leading to the teacher strongly intervening to generate results. It depends on the previous experiences of the students as to how open this task ought to be.

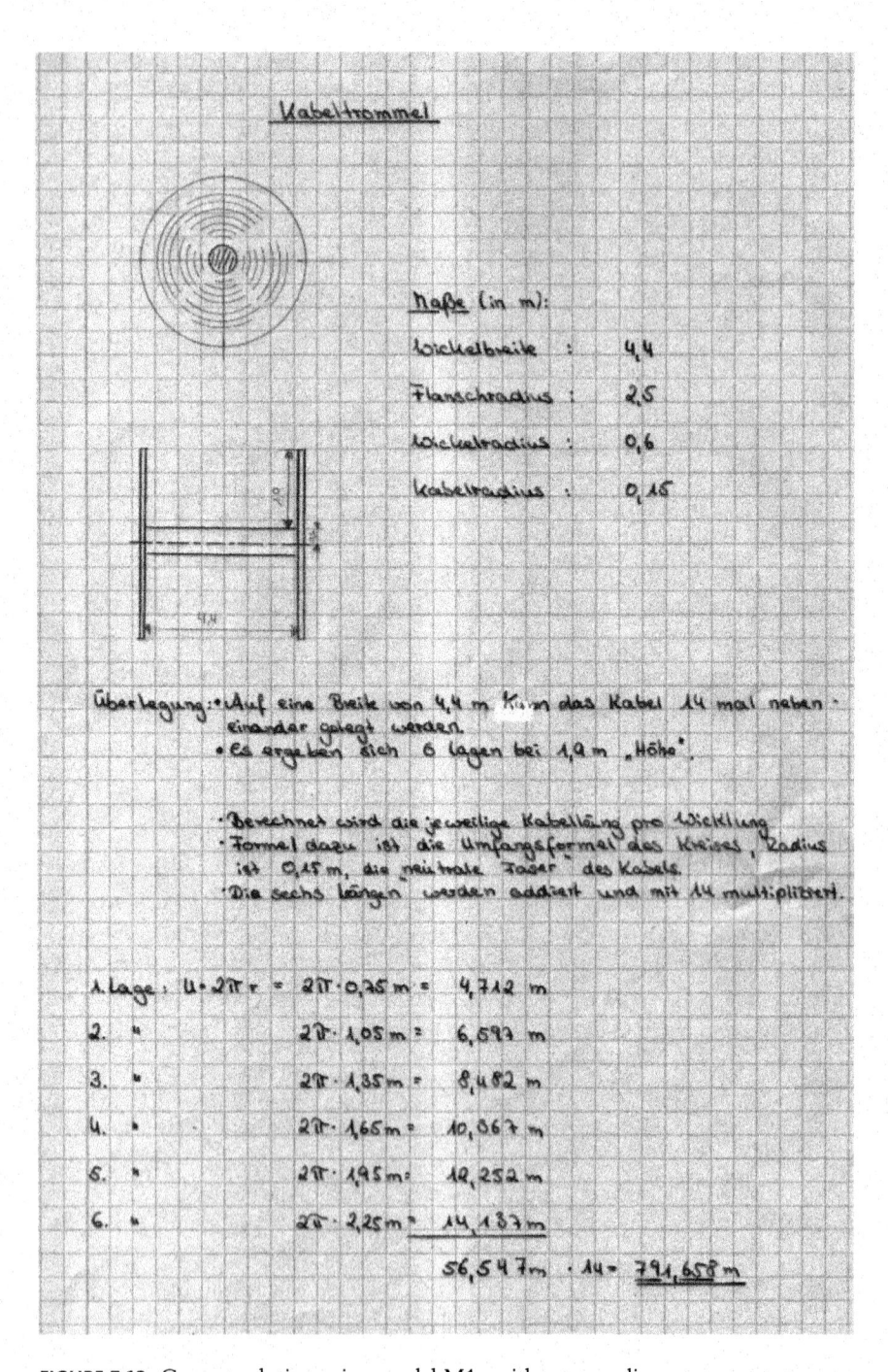

FIGURE 7.13 Correct solution using model M1a without rounding

References

Förster, F. (2018). Die Kabeltrommel "re-visited". In: H.-S. Siller, G. Greefrath & W. Blum (Eds.), *Neue Materialien für einen realitätsbezogenen Mathematikunterricht 4. 25 Jahre ISTRON-Gruppe – eine Best-of-Auswahl aus der ISTRON-Schriftenreihe* (pp. 319–330). Wiesbaden: Springer Spektrum.

Förster, F. & Herget, W. (2002). Die Kabeltrommel – Glatt gewickelt, gut entwickelt. In: *mathematik lehren* **113**, 48–52.

Förster, F. & Kaiser, G. (2010). The cable drum: Description of a challenging mathematical modelling example and a few experiences. In: B. Kaur & J. Dindyal (Eds.), *Mathematical Applications and Modelling, Yearbook 2010, AME (Association of Mathematics Education)* (pp. 276–299). Singapore: World Scientific.

Case 2: Painting a Porsche

The real-world problem and approximate solutions

One important factor in calculating the production cost of a car is the amount of paint and varnish needed to paint the surface. Normally, there are four layers of paint and varnish, which are applied on the surface at varying thickness (grounding, infill, basecoat, clear coat). To calculate the amount of paint needed, the surface area of the car has to be known. In addition, the cost for one layer depends on its thickness and on how many times the paint is applied.

How can the surface area of a car be determined? This question was the starting point of a teaching unit for the lower secondary level (grades 5 through 10), taught by Katja Maaß in Germany, mostly in grade 7 (see Maaß, 2004, 2006, 2018). Since Maaß was a teacher in Stuttgart, which is very close to Zuffenhausen where the Porsche is produced, she selected the Porsche 911 Carrera as an exemplary car (see Figure 7.14).

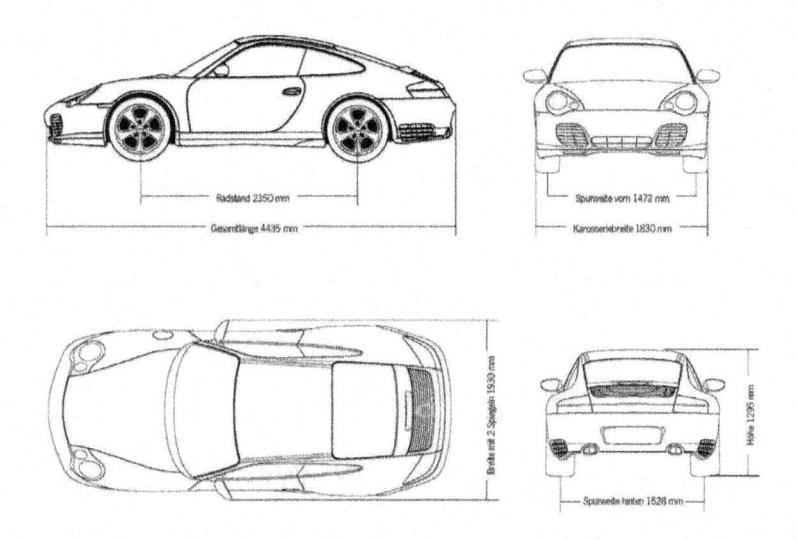

FIGURE 7.14 Measures of a Porsche 911

160 Cases of mathematical modelling

To determine the surface area, a mathematical (geometric) model of the car has to be constructed, which allows students to approximately calculate the surface area. Since the teaching unit was conceived for the lower secondary level, no advanced mathematical methods (such as modelling certain contours of the car by functions and using integrals for calculating corresponding areas) were to be used.

There are several possibilities for modelling such a car in an elementary way. The simplest geometric model consists of a right-angled parallelepiped (a cuboid) with the maximal outer measures of the car, which are (see Figure 7.14) approximately 4.5 m length, 1.9 m width and 1.3 m height. The surface of this solid (excluding the bottom) is:

$$2 \cdot (4.5 \text{ m} \cdot 1.3 \text{ m}) + 2 \cdot (1.9 \text{ m} \cdot 1.3 \text{ m}) + 4.5 \text{ m} \cdot 1.9 \text{ m} \approx 25.2 \text{ m}^2.$$

This is certainly an upper bound for the real surface area of the car and a very rough estimate. A more appropriate geometric model would be a right-angled parallelepiped which omits the windows and the wheels and therefore has a much smaller height, say 0.65 m instead of 1.3 m (half the height of the car). The resulting surface area (again excluding the bottom) is:

$$2 \cdot (4.5 \text{ m} \cdot 0.65 \text{ m}) + 2 \cdot (1.9 \text{ m} \cdot 0.65 \text{ m}) + 4.5 \text{ m} \cdot 1.9 \text{ m} \approx 16.9 \text{ m}^2.$$

We can find a better approximation if we decompose, on a picture, the surface of the Porsche (excluding all parts without paint such as windows, wheels or bumpers) into triangles or rectangles (see Figure 7.15 for the left side of the car), the real measures of which can be calculated by means of a suitable scale.

This calculation results in a surface area between approximately 13 m² and 17 m², depending on which plane figures are chosen, on the accuracy of measuring on the

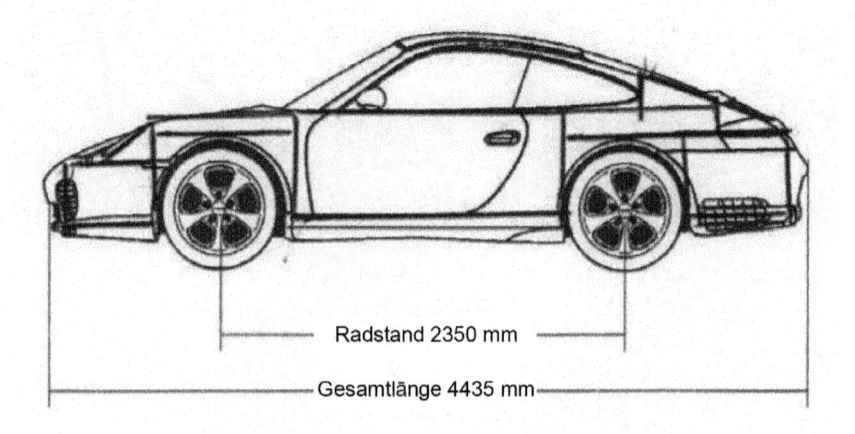

FIGURE 7.15 Decomposition of one side of a Porsche into triangles and rectangles

drawing and on the accuracy of the scale. The real surface area is, according to the producers (in a note to Katja Maaß), approximately 12 m².

The teaching unit "Surface of a Porsche"

In the following, we will present the teaching unit described in Maaß (2018). It was constructed primarily for grade 7 but can be implemented in other grades as well, supposing students have some previous knowledge of area. The unit started with presenting the problem to the students: "To calculate the costs of painting a Porsche, how can its surface area be determined?" Although lower secondary students are not car drivers themselves, this question is easy to understand and was interesting and motivating for them. The students quickly realised that some data are needed. The teacher supplied them with the data in Figure 7.14. A small difficulty was that all measures are in millimetres and have to be translated into metres because this is how lengths in the context of a car are primarily imagined. After a while, the students presented ideas of how to calculate an approximate value of the surface area (see the approaches above):

- Modelling the Porsche by a big parallelepiped in which it fits entirely;
- Using a smaller paralellepiped with only half the height of the big one;
- Covering the surface of the car with paper or cloths and measuring the area of this cover;
- Decomposing the surface into many small triangles and rectangles;
- Calling the Porsche company.

A whole-class discussion terminated with the decision not to pursue the idea of covering any further because no Porsche was available. The idea of calling the company was deputed to one of the students. The other three ideas were distributed to small groups (three to four students) to be further treated there ("big cuboid groups", "small cuboid groups", "decomposition groups"). In the course of the problem-solving process, several difficulties occured which caused the teacher to intervene, including the following:

- Some students in all groups had difficulties imagining the drawing spatially. A toy car provided by the teacher helped.
- Some students in the decomposition groups had troubles with the scaling. The teacher gave various hints, beginning by only giving the strategic prompt: "Imagine the relation between the picture and a real car as concretely as possible", continuing by referring to students' knowledge of maps and explaining, if needed, the relation between the measures on the picture and in reality.

After the group work, some students presented their calculations and findings to the whole class. The three approaches were compared, and it became clear that there is no right solution but a variety of approximations. Since some groups had unreasonably accurate results ("13.2375 m^2"), the teacher explained why it does not make sense to give so many decimal points. It turned out that the results of the small cuboid groups and the decomposition groups were quite similar, so the students agreed that the small cuboid model is the most suitable one because it is both sufficiently easy and sufficiently precise. Eventually, the students wanted to know how the Porsche company actually determines this surface area. The teacher explained that a CAD programme is used which is based on similar decomposition methods.

Evaluating the teaching unit

The teaching unit described in Maaß (2018) was completed with a regular class test that was adjusted to the unit. The teacher knew that the students will only take seriously what is assessed and emphasised that all elements of the unit might be part of the test. The test actually used by Katja Maaß in her grade 7 classes consisted of three tasks (see Figure 7.16), only one of which is in direct continuation of the teaching unit, whereas the two other tasks deliberately required some transfer.

Task 1 was a direct translation of the considerations and calculations made in the Porsche context into the context of a Mercedes: "Describe various possibilities to calculate the surface area of the body of a Mercedes class A. Compare the approaches and the expected results. Calculate the surface area according to a method chosen by you." The students were expected to essentially repeat what had been done in the unit and to find a result in the range of 13 m^2 to 17 m^2. In task 2, the students had to calculate the difference in the costs of covering a Parquet floor with two different materials, luxury maple ("Ahorn-select") or beech ("Buche"). They had to first select the necessary information from the given price table and then calculate the area based on the given map, where one missing length had to be calculated using the underlying scale. Finally, they had to calculate the resulting difference in costs. Task 3 contained a problematic proportional calculation in the context of running ("10 km in 40 minutes means 42 km in 168 minutes") and the students were asked, "What is your opinion on this?". The students were expected to activate their real-world knowledge of running and argue that proportional reasoning will, most probably, be inappropriate in this context.

For the marking of this class test and of modelling tasks in general, Maaß developed the following scheme which her students were informed about in advance:

1 *Construction of the real model* (see Figure 2.8) (Do the assumptions make sense, are the simplifications appropriate?): up to 10 points;
2 *Use of mathematics* (Are the variables and relations mathematised correctly, is the mathematical notation appropriate, are mathematical knowledge and problem solving strategies applied correctly, is the solution mathematically correct?): up to 15 points;
3 *Interpretation of the solution* (Is the mathematical solution interpreted correctly in the real world?): up to 5 points;
4 *Critical reflection* (Are all relevant aspects considered, are reference values used?): up to 10 points;

Klassenarbeit

Aufgabe 1
Beschreibe mehrere Vorgehensweisen, um die Oberfläche der Karosserie des Mercedes der A-Klasse zu berechnen. Vergleiche die Vorgehensweisen und die zu erwartenden Ergebnisse. Berechne die Oberfläche nach einer von dir gewählten Methode.

Aufgabe 2
Herr Schulze möchte seine neue Wohnung mit Parkett auslegen lassen, kann sich aber nicht zwischen Ahorn-select und Buche entscheiden. Ahorn-select gefällt ihm besser, ist aber teurer. Berechne den ungefähren Preisunterschied zwischen dem Auslegen der Wohnung mit Ahorn-select und Buche.

Parkett-Preise
Ahorn	79,90 €/qm
Ahorn-select	89,90 €/qm
Buche	59,90 €/qm
Buche–rustikal	49,90 €/qm
Eiche	69,90 €/qm
Arbeitslohn	20–30 €/qm
Fußleisten	8–15 €/m + Montage

Hinzu kommen die Kosten für die Übergangsleisten zwischen den Zimmern, mehr Arbeitslohn für schwierige Stellen, Anfahrt.

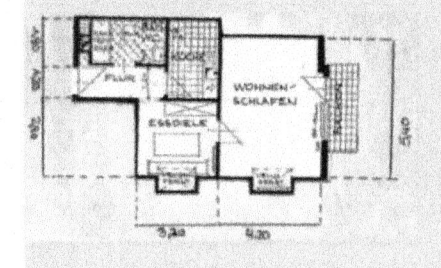

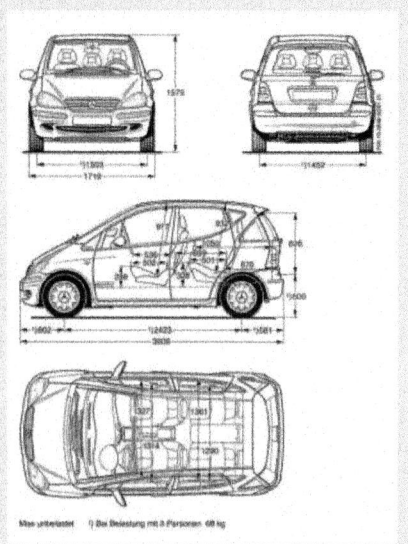

Aufgabe 3
Peter ist ein begeisterter Jogger. Er läuft 10 km in 40 min. Nun möchte er einen Marathon laufen. Er rechnet:

Entfernung	Zeit
10 km	40 min
1 km	4 min
42 km	168 min = 2 Std. 48 min

Was meinst du dazu?

FIGURE 7.16 The class test after the Porsche unit

5 *Documentation of the way of proceeding* (Are the steps of the solution process described and explained?): up to 15 points; and
6 *Goal-directedness* of the solution (Does the student proceed in a goal-oriented way or does he/she get lost in details?): up to 5 points.

Altogether, an answer to this task could receive maximum of 60 points. This scheme was used not only for this unit but generally in Maaß's teaching (see Maaß, 2004, for the entire empirical study based on her own teaching in a German *Gymnasium*).

The Porsche unit was conceived as an introductory unit in modelling, which means the learners were not supposed to have had modelling experiences before. Maaß reports that most of her grade 7 students had no pre-knowledge, but the results of task 1 in a class test proved that most students had, after this unit, no difficulties with typical modelling steps such as identifying relevant data, making assumptions, choosing models, rounding off results or comparing the outcomes of different models, providing the contexts were familiar. The mistakes made in task 1 were rather due to insufficient mathematical knowledge and skills, such as the use of a wrong formula ("$V = a \cdot b$" for the volume of a cuboid), a wrong conversion of units (1830 mm = 18.3 cm") or an inadequate conception of space. The results of task 2 were even more encouraging since here, too, relevant data had to be extracted and assumptions had to be made but now in an unfamiliar context. Most mistakes in task 3 were made when some students reverted to their usual behaviour when solving word problems by applying a proportional calculation without thinking about the context and without validating their result. This behaviour changed only gradually in the course of subsequent modelling units. An unintended effect of the Porsche unit was that some students took the following message away from the work: "Actually, you cannot make mistakes because nobody can control whether your solution is correct or not." (Maaß, 2006, p. 135). Maaß's study also showed how stable students' beliefs are and that these can only be changed in long-term processes. Some students were able to cope with modelling tasks from the beginning but were not interested in real-world connections (Maaß calls them "uninterested modellers") or were not willing to engage in these ("reality-distant modellers"). Two other types of students were those with positive attitudes towards real-world connections combined with positive attitudes towards mathematics ("reflective modellers") or with negative attitudes towards mathematics ("mathematics-distant modellers"). Maaß observed strong correlations between attitudes and performance, already in the first class test discussed above (for more details, see Maaß, 2004, 2006).

References

Maaß, K. (2004). *Mathematisches Modellieren im Unterricht*. Bad Salzdetfurth: Franzbecker.

Maaß, K. (2006). What are modelling competencies? In: *ZDM Mathematics Education* **38**(2), 113–142.

Maaß, K. (2018). Der Porsche 911. Mathematisches Modellieren für Anfänger. In: H.-S. Siller, G. Greefrath & W. Blum (Eds.), *Neue Materialien für einen realitätsbezogenen Mathematikunterricht 4. 25 Jahre ISTRON-Gruppe – eine Best-of-Auswahl aus der ISTRON-Schriftenreihe* (pp. 285–292). Wiesbaden: Springer Spektrum.

7.3 Curricular programmes of mathematical modelling

In this section, we present a university curriculum and a school curriculum with a substantial modelling content, implemented over many years in the respective educational settings. These cases are exemplified by concrete excerpts from these curricula. For reasons of space, we abstain from supplying further details of these programmes.

Case 3: Mathematical modelling at Roskilde University

The modelling course BASE

Since 1972, when Roskilde University admitted its very first students, mathematical modelling has been a key component in the mathematics and science programmes at this university. Students' independent construction of mathematical models in various domains of science, society, technology and culture, as well as their critical analysis of the foundation and properties of existing models, have been on the agenda of the programmes for almost half a century in a multitude of ways. The predominant avenue for students' constructive and analytical modelling activities has always been problem-oriented project work, in which students in small groups (two to eight participants) work, under guidance by faculty supervisors, half of the time in a semester to pose and answer various questions in, about or by means of mathematical modelling (Blomhøj & Kjeldsen, 2011, 2018).

This is not the place to provide a comprehensive presentation of the place and role of modelling at Roskilde University with all its facets (for a general account of the Roskilde model, see Niss, 2001). Instead, we shall focus on one course on mathematical modelling that was given as an optional part of a special two-year programme across the sciences, which constituted the compulsory introduction to all studies in (natural) science and mathematics at Roskilde, until the university was forced to change its programmes to conform with the so-called Bologna structure of university studies in the European Union.

The course under consideration, BASE (a Danish acronym referring to basic analysis, simulation and experiments) was developed and first implemented in the academic years 1999–2000 and 2000–2001 and was offered for the last time in the academic year 2009–2010, when the study structure was fundamentally changed. BASE has been subject to presentations and analyses published in journal papers and book chapters, including Blomhøj et al. (2001), Ottesen (2001), and Blomhøj and Kjeldsen (2009). The exposition of BASE below is based on Blomhøj et al., (2001) and Blomhøj and Kjeldsen (2010). The modifications of the course during the remainder of its lifetime were marginal and will not be dealt with here.

The purpose of the course was stated as follows (Blomhøj et al., 2001, p. 11):

> The course is meant to support the development of model[ling] competency with the students so that they become able to construct, apply, analyse and criticise mathematical models in simple problem situations.
>
> *(Our translation)*

The target audience of the course consisted of newcomers to the science and mathematics studies at Roskilde University, especially students with relatively weak backgrounds in mathematics from upper secondary school. It also served the purpose of supplying these students with upper secondary mathematical prerequisites cast in terms of mathematical models and modelling so that they could cope with the general mathematical demands of studies in science and mathematics, which at the time included a compulsory course component in mathematics amounting to a fourth of the workload of one full year of study. In large part, the primary purpose of the course was to use modelling as a vehicle for learning mathematics needed for the study of science (and mathematics) at the university level, whereas mathematical modelling as an independent goal played a secondary yet still significant role. Even though BASE was optional, it was the recommended and obvious choice for students with a frail motivation for mathematical studies and a weak mathematics background. Students with a stronger mathematical background from upper secondary school were advised to take courses in linear algebra, in calculus and analysis, and in statistics, unless they explicitly wanted to revisit upper secondary mathematics from a models and modelling perspective.

The duration of the course was two semesters with two 3-hour sessions per week. The bulk of the course (roughly 75%) was devoted to students' work on so-called mini modelling projects, where groups of three to four students collaborated on constructing, analysing, interpreting and critiquing a mathematical model. The themes and topics of the mini-projects were given by the course directors, which was also the case with the written introduction to each project included in the notes for the course. Student groups' written reports on their work were submitted to the course directors – who were also the teachers – for correction, commentary and assessment. Each student was involved in conducting 5–6 mini-projects throughout the course but also acted as a peer assessor of another 5–6 mini-projects done by fellow students, such that they became closely acquainted with a total of 10–12 such projects. The project groups decided themselves how to distribute the mini-projects among themselves, whereas the distribution of peer review assignments was decided by the teachers. One mini-project typically took 12 course hours plus 8–10 homework hours to complete.

In addition to project work, students also attended lectures and introductions to the mini-projects and to mathematical concepts and methods, including numerical methods, as part of the course. Moreover, students were given many exercises and problems focusing on mathematical topics pertaining to the themes under consideration. The teaching materials were lecture notes written by the course directors themselves (Blomhøj et al., 2001a, 2001b).

Modelling situations in BASE

As mentioned, the modelling situations to be dealt with in the mini-projects were chosen and presented in written form by the course directors and were grouped into three parts. The first two parts, which were placed in the first of the two semesters, focused on modelling by way of special functions (the first part), including linear functions, linear regression, exponential functions, and power functions as well as mathematical transformations of such functions, and (in the second part) on modelling involving differential and integral calculus. In the first two parts of the course, students knew that they were supposed to make use of functional modelling, but usually it was not specified which particular functions from the upper secondary repository of functions might be relevant in the given context. The third part, which made up the entire second semester, dealt with modelling of dynamic phenomena by means of differential equations, especially compartment modelling, and systems of differential equations, including phase-plane analysis of the properties of their solutions. The modelling situations were sequenced so as to involve progression such that both the level of the mathematics involved and the complexity of the modelling tasks increased throughout the course. Throughout the course, the tasks were designed and presented in such a way that students had to spend most of their time on mathematisation, mathematical treatment of the model obtained, as well as on de-mathematisation and on the basics of outcome validation. Less attention was supposed to be paid to other components of the modelling cycle, such as pre-mathematisation and model evaluation. The course included an introduction to the relatively advanced digital software MatLab, which was meant to be useful for mathematical treatment in modelling contexts where the mathematical problems that arose could not be solved analytically but must be solved numerically. The modelling situations from which the mini-projects were to be conducted during the first two years of implementation of the course were the following; in the first semester, six modelling situations were on the agenda:

1   Monod's growth chamber experiment,
    - How does a bacterial strain grow?
    - How does the activity of a cell depend on a ligand?

2   Metabolism and mass
    - What is the relationship between metabolism and surface area of warm-blooded animals?

3   Biological diversity and the size of an island
    - What is the relationship between the number of animal species on an island and its size?

4   Systematic features of our planetary system
    - What are the relationships between the distances of planets to the Sun, their place numbers and their orbitals?

5   The age of the Earth
    - How old is the Earth?

6   "10 = 44" – a traffic speed campaign
    - A traffic speed campaign claimed that if two cars driving next to each other in parallel lanes, one at a speed of 50 km/h, the other one at 60 km/h, brake at the same time to avoid an object a small distance away in

front of them, the car at 50km/h can just manage to stop, whereas the other car will hit the object at a speed of 44 km/h. Is the claim in the campaign correct?

In the second semester, eight modelling situations were presented to the students:

1. Dramatic population growth
   - What is the development of the world's human population?
2. The cormorant population in Denmark
   - What is the development of the Danish cormorant population?
3. Drug dosages
   - What is a good medication plan for a person of 50 kg suffering from asthma?
4. Radio therapy against cancer tumours
   - What should a dosage plan for radio therapy look like?
5. Anaesthesia
   - How should anaesthetic drugs be dosed during a surgical operation?
6. Predator-prey systems
   - What are the basic dynamics in the interplay between predator animals and their prey?
7. Modelling of the development of epidemics
   - What governs the dynamics of an epidemic?
8. Modelling of gonorrhoea
   - What governs the prevalence of gonorrhoea?

For each modelling situation, a written presentation of the background of the task (sometimes a rather lengthy one), together with specific information and numerical data, when relevant, was given to students. Also, the main questions to be answered in the mini-project were posed in the stimulus text, and typically a number of hints and scaffolding questions were offered.

The summative assessment criteria for the course had three components. The most important criterion was timely submission of all mini-project reports. Two 15-minute individual oral interviews, one at the end of each semester, on two mini-project reports drawn from the pool at random constituted the second and the third assessment component. Students received a "passed" or "failed" mark at the end of the course. In addition to this summative assessment, the teachers of the course provided formative feedback to students, both individually and in groups, at crucial points during the course, especially when giving feedback on completed mini-project reports. This feedback was focused on assisting students in making progress in the remainder of the course.

An example: modelling the CO_2 balance of a lake

This example from a later version of the course, presented in Blomhøj and Kjeldsen (2010), deals with the way in which students' work on a modelling problem in a mini-project served the purpose of paving the way for understanding the concept of the (definite and indefinite) integral and the fundamental theorem of calculus, i.e., the relationships between a function, its definite integral and its antiderivative, especially when the function designates the rate of change of another – initially unknown – function. Prior to the mini-project, the course directors had experienced a mismatch in students' concept images of the antiderivative and of the definite integral, especially in settings where the function did not have an analytically well-defined anti-derivative. They wanted to counteract this mismatch in a modelling setting that called for numerical integration, that is, for reconstructing an unknown function by means of its known derivative, especially for functions initially defined only on a discrete set of points. The setting was the CO_2 balance of a lake. In a lake, during the day plants use CO_2 in the process of photosynthesis, while they produce CO_2 during the night. Animals in the lake produce CO_2 continuously due to respiration. Biologists are interested to know the net rate of change for such a lake during a 24-hour period. The students were given data representing the rate of change of CO_2 (mmol/litre)/hour in the lake every 40 minutes during 24 hours after dawn, corresponding to 36 data points. The initial value of the CO_2 content at dawn was 2600 mmol. During the first 12 hours after dawn, the rates of change were all negative, whereas they were positive during the next 12 hours, and they were 0 at 12 and 24 hours after dawn.

The opening question given to students was: "What can be concluded from the data material about the sign of the rate of change of CO_2 over the 24-hour period, and what information does that give about the life in the lake?"(p. 576). The teachers supervising the groups and giving guidance along the road from time to time came up with auxiliary questions, for example:

> When will the CO_2 content be at its lowest, and how much CO_2 will be in the water when that happens? Is the lake in equilibrium with regard to the CO_2 content? How can this question be decided graphically? How much CO_2 was released to the water during the 12 hours at night and how much was removed during the 12 hours of daytime?
>
> *(p. 577)*

Students were led to answer these questions by numerical integration performed by hand, by way of MatLab or by using Excel. The teachers further made pedagogical observations during the process and concluded that despite several hurdles and challenges encountered on the way, students did indeed markedly develop and consolidate their concept of the integral and their understanding of the integral got connected to the definition of the concept. They also experienced the usefulness of simple hands-on numerical integration (p. 579), even though they initially discarded this as "childish".

Final remarks

It is important to underline, once again, that the course BASE was never the only avenue for mathematical modelling in the mathematics and science programme at Roskilde University. Problem-oriented project work involving mathematical modelling undertaken by small groups of students under guidance by faculty supervisors was always a key component of the programme. The purpose of such modelling activities is to foster modelling and modelling competencies as an independent goal rather than as a vehicle to other ends, such as learning mathematical concepts, methods and results. Even though, for structural reasons, BASE was withdrawn from the study programme in 2010, mathematical modelling activities continue to form a crucial part of the mathematics and science programme (Blomhøj & Kjeldsen, 2018). Also, these modelling projects involve all the phases of the modelling cycle, not only those in focus in BASE.

References

Blomhøj, M., Jensen, T.H., Kjeldsen, T.H. & Ottesen, J. (2001). *Matematisk modellering ved den naturvidenskabelige basisuddannelse – udviklingen af et kursus* (Mathematical Modelling in the Basic Science Programme [at Roskilde University]: Developing a Course). Tekster fra IMFUFA, nr. 402: Roskilde: IMFUFA/Roskilde University.

Blomhøj, M. & Kjeldsen, T.H. (2009). Project organised science studies at university level: Exemplarity and interdisciplinarity. In: *ZDM: The International Journal on Mathematics Education* 41(1–2), 183–198.

Blomhøj, M. & Kjeldsen, T.H. (2010). Learning mathematics through modeling: The case of the integral concept. In: B. Sriraman, L. Haapasalo, B.D. Søndergaard, G. Palsdottír & S. Goodchild (Eds.), *The First Sourcebooks on Nordic Research in Mathematics Education* (pp. 569–581). Charlotte, NC: Information Age Publishing.

Blomhøj, M. & Kjeldsen, T.H. (2011). Students' reflections in mathematical modelling projects. In: G. Kaiser, W. Blum, R. Borromeo Ferri & G. Stillman (Eds.), *Trends in the Teaching and Learning of mathematical Modelling* (pp. 385–395). Dordrecht: Springer.

Blomhøj, M. & Kjeldsen, T.H. (2018). Interdisciplinary problem oriented project work: A learning environment for mathematical modelling. In: S. Schukajlow & W. Blum (Hrsg.), *Evaluierte Lernumgebungen zum Modellieren* (pp. 11–29). Wiesbaden: Springer Spektrum.

Blomhøj, M, Kjeldsen, T.H. & Ottesen, J. (2001a). *BASE Note 1, August 2000*. Roskilde: Nat.Bas, RUC.

Blomhøj, M, Kjeldsen, T.H. & Ottesen, J. (2001b). *BASE Note 2, November 2000*. Roskilde: Nat.Bas, RUC.

Niss, M. (2001). University mathematics based on problem-oriented student projects: 25 years of experience with the Roskilde model. In: D. Holton (Ed.), *The Teaching and Learning og Mathematics at University Level: An ICMI Study* (pp. 153–165). Dordrecht: Kluwer Academic Publishers.

Ottesen, J. (2001). Do not ask what mathematics can do for modelling: Ask what modelling can do for mathematics! In: D. Holton (Ed.), *The Teaching and Learning og Mathematics at University Level: An ICMI Study* (pp. 335–346). Dordrecht: Kluwer Academic Publishers.

Case 4: Modelling in Australia – Queensland

Modelling in some Australian curricula

For several decades, states in Australia have paid varying amounts of attention to mathematical models and mathematical modelling in school curricula. Education in Australia is a state and territory issue rather than a national one, which means that there is considerable diversity across the country, even though there are tendencies towards an increased centralisation of curricula. Two Australian states have been protagonists in placing applied problem solving, models and modelling on their curricular agenda: Victoria and Queensland. A leading figure in this development is Peter L. Galbraith who for decades has been active in research and development in both states and has been a mentor for several Australian researchers in the field since the 1990s.

After some precursors a few years before, Victoria introduced applied problem solving and mathematical modelling into an innovative curriculum for upper secondary mathematics education already in 1988. Thus, in Victoria Board of Curriculum and Assessment (1988) one can read (quoted from Stillman, 2007, p. 498) about the incorporation of "problem solving and modelling activities [. . .] intended to provide students with experience in using their mathematical knowledge in creative ways to solve non-routine problems" (p. 24) and about the engagement of students in investigative projects such as "an extended mathematical modelling exercise to solve a real-world problem" (p. 27). This curriculum was first implemented in a few schools, but it was fully implemented in Victoria for year 11 students in 1990 and for year 12 students in 1991 (Stephens & Money, 1993; Stillman, 2007). In the first years of implementation, the so-called CATs, Common Assessment Tasks, of this curriculum involved a marked component of problem solving and mathematical modelling. However, during the 1990s, this component became more and more reduced, a fact that entailed a reduced emphasis on modelling in Victorian upper secondary schools at large.

In Queensland, which is the focus of this case, applications, models, applied problem solving and modelling have a rather long and sustained history in mathematics education. As far as senior (upper) secondary curricula (years 11 and 12) are concerned, this development gained momentum after the introduction of senior secondary trial/pilot syllabuses in 1989 (Queensland Board of Senior Secondary School Studies, 1989a, 1989b) and the subsequent implementation of these in a limited number of trial/pilot schools from 1990–91 on (Stillman & Galbraith, 2009, 2011). These curricula emphasised all the key aspects of mathematical modelling as manifested in the full modelling cycle. Even though the specific formulations adopted in curricula and syllabuses have changed over time, the spirit and thrust have remained stable since the early days.

Recently, Australia has become a key player in the International Mathematical Modelling Challenge for upper secondary students through The Australian Council for Educational Research (ACER) (for more details, see section 5.6), thus consolidating the long-established prominent role of mathematical modelling in the country.

Modelling and assessment in Queensland

Until recently, Queensland was the only state in Australia in which assessment is entirely school based (Stillman & Galbraith, 2011). According to Stillman and Galbraith (2009):

> The Queensland system of school based assessment means that the production of assessment tasks and the award of levels of achievement is in the hands firstly of individual schools, with panels at district and state levels performing critical reviewing roles to assure comparability of outcomes across schools and regions. In keeping with the school based nature of the Queensland context, individual schools and teachers design individual work programs (including assessment tasks) under the syllabus umbrella. Hence a key element is the translation of the general objectives into specific criteria for the teaching, learning, and assessment of school based activity.
>
> *(p. 517)*

This means that to come closer to what actually happens in schools, it is necessary to zoom in on the workings and the tasks of the individual schools. Stillman and Galbraith (2009), in an empirical study of 23 teachers and "curriculum figures" from across the state, found quite a diversity of implementation across schools; they also found that even though teachers in general agreed that modelling was established in classrooms (op. cit., p. 520), several of them did not see a significant distinction between applications of mathematics and mathematical modelling. The study further found that the role of the mediating panels mentioned in the quote above was to gradually and softly nudge schools and teachers to embrace all facets of mathematical modelling more fully.

To illustrate how one secondary school in Queensland, Ormiston College, interpreted and implemented the modelling parts of the curriculum, we present some examples of modelling assessment tasks given to students in this particular school under the leadership of Ian Thomson.

In 2011, year 12 students in Mathematics B: Extended Modelling and Problem Solving (the subject studied by students intending to undertake serious mathematics studies at the tertiary level; see Stillman & Galbraith, 2009, p. 516) were given two weeks to investigate, analyse and report on differences in daylight between Melbourne and Brisbane, with a special emphasis on wintertime.

In 2013, year 12 students taking Mathematics B were presented with data about the braking and stopping of trains and cars and were asked to investigate a range of braking situations by way of mathematical modelling involving calculus.

In 2014, year 12 Mathematics B students were requested to make use of the so-called Lorenz Curve and the Gini Coefficient, as well of calculus, to investigate aspects of economic inequality in different countries.

In 2015, year 12 Mathematics B students were shown a video clip of a top basketball player with a remarkable record of achievement in accurate shots. Taking inspiration from the video, they were asked to investigate and model the trajectories of basketballs in shots and explain why the player at issue had an advantageous technique. Here, too, two weeks were given to conduct the investigation and report the outcomes.

Mathematics C is the most challenging mathematics course offered to senior secondary students in Queensland. In 2015, the year 11 students in the school under consideration had to investigate and utilise methods of encryption and decoding to model secret codes in several specific examples.

In conformity with the official Queensland syllabus objectives, over the years the written reports of students in this school were assessed based on three categories of criteria: "knowledge and procedures", "modelling and problem solving", and "communication and justification".

From 2019 onwards, Queensland, through its Queensland Curriculum and Assessment Authority (QCAA), implemented a new mathematics syllabus for senior (upper secondary) mathematics (www.qcaa.qld/edu.au/downloads/senior/). Mathematics is divided into four different subjects/courses: Essential Mathematics, General Mathematics, Mathematical Methods and Specialist Mathematics. A compulsory summative external examination comprising 50% of the total assessment has been introduced for all four courses. In addition, each course involves a coursework-based problem solving and modelling task worth 20% of the marks earned. The task has to include the following components, which also form the basis of the assessment described in an instrument-specific marking guide that teachers have to employ: Formulate (3–4 marks), Solve (6–7 marks), Evaluate and Verify (4–5 marks) and Communicate (3–4 marks). Examples given in this syllabus include: Make a recommendation for the most appropriate type of water tank for installation in a house, considering the amount of collectable rainfall and water usage (Essential Mathematics); investigate and compare students' attitudes to environmental sustainability issues, including pollution, today and in the 1990s (General Mathematics); investigate motion sickness on a Ferris wheel using the vertical velocity and acceleration of a Ferris wheel car (Mathematical Methods); write a report showing how matrices can be used to predict the eventual winner of a competition (Specialist Mathematics).

It is interesting to note that modelling in Queensland is, to a large extent, fostered and promoted by the rules and criteria of assessment. By setting rather specific standards for problem solving and modelling tasks and the assessment of them, the curriculum authorities indirectly drive teachers to undertake mathematical modelling activities. In other words, this is an example of *assessment driven curriculum reform* pertaining to mathematical modelling.

References

Queensland Board of Senior Secondary School Studies (1989a). *Trial/Pilot Senior Syllabus in Mathematics B*. Brisbane: Author.

Queensland Board of Senior Secondary School Studies (1989b). *Trial/Pilot Senior Syllabus in Mathematics C*. Brisbane: Author.

Stephens, M. & Money, R. (1993). New developments in senior secondary assessment in Australia. In: M. Niss (Ed.), *Cases of Assessment in Mathematics Education: An ICMI Study* (pp. 155–171). Dordrecht: Kluwer.

Stillman, G. (2007). Implementation case study: Sustaining curriculum change. In: W. Blum, P.L. Galbraith, H-W. Henn & M. Niss (Eds.), *Applications and Modelling in Mathematics Education: The 14th ICMI Study* (pp. 497–502). New York, NY: Springer.

Stillman, G. & Galbraith, P. (2009). Softly, softly: Curriculum change in applications and modelling in the senior secondary curriculum in Queensland. In: R. Hunter, B. Bicknell & T. Burgess (Eds.), *Crossing Divides: Proceedings of the 32nd Annual Conference of the Mathematics Education Research Group of Australasia* (Vol. 2, pp. 515–522). Palmerston North, NZ: MERGA.

Stillman, G. & Galbraith, P. (2011). Evolution of applications and modelling in a senior secondary curriculum. In: G. Kaiser, W. Blum, R. Borromeo Ferri & G. Stillman (Eds.), *Trends in Teaching and Learning of Mathematical Modelling: ICTMA 14* (pp. 689–697). New York, NY: Springer.

Victorian Board of Curriculum and Assessment (1988). *Mathematics Study Design: Provisionally Accredited*. Melbourne, Australia: Victoria Board of Curriculum and Assessment.

7.4 Modelling materials

This final section presents materials that have been developed for teaching units on mathematical modelling in school or university classrooms. We illustrate the first two cases (both for the secondary school level) by concrete examples contained in those materials.

Case 5: The Shell Centre at Nottingham University

The Shell Centre for Mathematical Education at Nottingham University, UK, was established in 1967. From its early days, it was committed to developing, testing and disseminating carefully produced and well-written teaching and learning materials for mathematics education, with a focus on applied mathematical problem solving and modelling, especially with regard to everyday problems of interest to primary and secondary school students. The materials consist of stimulus units and modules for teaching, student exploration, formative and summative assessment, as well as advice and guidelines for teachers. The modelling focus became particularly visible when Hugh Burkhardt (for more historical information, see Burkhardt, 2018) was appointed Director of the Centre in 1976 and was amplified when Malcolm Swan joined the Centre in 1979. This focus was further exacerbated in follow-up work to the highly influential so-called Cockcroft report (1982), which gave momentum to mathematics educators' attention to the notion and significance of "numeracy" throughout the world. In section 5.6, we mentioned the *Numeracy Through Problem Solving Series* which came out in direct response to the Cockcroft report. Later, the Centre also took an interest in mathematical literacy as developed in PISA and elsewhere (Stacey & Turner, 2015)

The Centre, which celebrated its golden jubilee in 2017, is now part of the Centre for Research into Mathematics Education under the School of Education at Nottingham University, for which Geoff Wake is the current Centre Convenor. It continues to work along the lines mentioned, oftentimes in collaborative projects with parties in other countries and in recent years with particular emphasis on materials for assessment in mathematics in general and problem solving, modelling and mathematical literacy in particular, as well as problem-based learning in science and mathematics. It is characteristic that faculty at the Centre have always accompanied their development work by theoretical and empirical research papers and books.

Among the materials and publications that had a major impact on the development of mathematics teaching as well as on students' learning activities were *Problems With Patterns and Numbers* (Shell Centre, 1984), presenting a rich variety of non-routine problem solving tasks for O-level (O standing for "ordinary") students ages 13 to 16, most of which were designed as one- or two-week modules. The following year, the Centre published, again for 13- to 16-year-old students, *The Language of Functions and Graphs* (Shell Centre, 1985) in two units, A and B, where Unit B was devoted to presenting a number of two-week modelling tasks involving functions and graphs. In 1989, the Centre published the first edition of *Extended Tasks for GSCE Mathematics* (Shell Centre, 1989) for school-based assessment, in which the Teacher Guide sets the stage and provides the overall framework for the tasks. Many modelling examples can be found in the collection of lessons for grade 6 onwards produced in the Mathematics Assessment Project; see www.map.mathshell.org/lessons.php. All modules contain materials for teaching and for assessment. Most of the materials mentioned can be downloaded for free from the Centre's webpage (www.mathshell.com) or can be obtained by writing to the Centre.

Particularly influential far beyond the British mathematics education scene was *The Language of Functions and Graphs*. Qualitative graphs like those in Figure 7.17

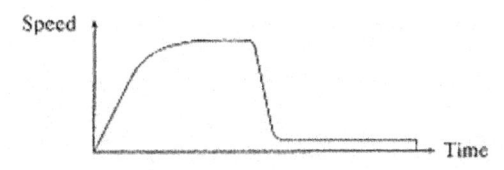

FIGURE 7.17 Example from *The Language of Functions and Graphs*

were seldom found in any materials around the world before the Shell Centre published their units and presented the materials and experiences at international conferences.

Both the real-world interpretation of given qualitative graphs and the creation of such graphs to fit given real-world situations are now part of mathematics textbooks in many countries. The examples in this module cover a wide range of areas including sports, camping, traffic, growth, tides and many others.

References

Burkhardt, H. (2018). Why teach modelling: A 50 year study. In: *ZDM: The International Journal on Mathematics Education* **50**(1 + 2), 61–75.

Report of the Committee of Inquiry into the Teaching of Mathematics under the Chairmanship of Dr. W.H. Cockroft (1982). *Mathematics Counts*. London: Her Majesty's Stationery Office.

The Shell Centre (1984). *Problems with Patterns and Numbers*. Nottingham: Shell Centre Publications.

The Shell Centre (1985). *The Language of Functions and Graphs*. Nottingham: Shell Centre Publications.

The Shell Centre (1989). *Extended Tasks for GSCE Mathematics*. Nottingham: Shell Centre Publications.

Stacey, K. & Turner, R. (Eds.) (2015). *Assessing Mathematical Literacy: The PISA Experience*. New York: Springer.

Case 6: Best of Istron

The German Istron group

In section 5.6, the German Istron[2] group was mentioned. This group was founded in 1991 with the aim of improving the teaching of mathematics at all school levels through the inclusion of connections to the real world. At that time, applications and modelling were seldom found in German mathematics classrooms, much more seldom than today where modelling is one of six mathematical competencies which, according to the German education standards, are compulsory for mathematics teaching and assessment from grade 1 on. The group consisted, from the beginning, of teachers and researchers in schools and universities in Germany and Austria who had the common intention of fostering the inclusion of mathematical modelling and applications of mathematics in everyday classrooms and in all kinds of assessment. At present, the group consists of around 70 people. There is no formal membership. The group is organised by two elected members and meets twice a year. Since 1997, once a year the meeting is combined with a whole-day in-service teacher training event where Istron members offer various lectures and workshops on mathematical modelling for local teachers. These meetings and courses often take place in German or Austrian universities and are organised by local Istron members.

Another activity of the Istron group is the editing and publishing of a series of books for teachers under the title *Materialien für einen realitätsbezogenen Mathematikunterricht (Materials for Reality-Oriented Mathematics Teaching)*. The first volume was published in 1993; there are now 24 volumes, 18 of which were published by Franzbecker, whereas the last 6 ones, called *Neue Materialien für . . . (New Materials for . . .)*, were published by Springer. All volumes contain articles presenting modelling examples and reports on classroom experiences with modelling. The articles are written for mathematics teachers as the audience, with the aim of supplying teachers with ideas and materials which they can directly use in their everyday teaching practice. Some of the volumes have an overarching topic such as the use of digital tools in modelling or modelling examples for the lower secondary level, but most volumes are just a collection of interesting reports from everyday practice of mathematics instruction in schools. The webpage www.istron.mathematik.uni-wuerzburg.de/istron/index.html@p=1033.html contains the tables of contents of all Istron books as well as abstracts of all contributions. Nearly 300 articles have been published in this series.

The Best-of-Istron volume

On the twenty-fifth anniversary of the Istron group in 2016, the group decided to compose a book containing a "Best of" selection of all 18 Istron books published by Franzbecker. The criteria for selecting the articles for this volume were that these articles have, according to teachers' reports, proven particularly useful in school practice and that the editors regard them as particularly typical of the "spirit" of Istron. Altogether, the modelling examples in these articles were to cover a sufficiently broad spectrum of school levels, mathematical topics and real-world contexts. In addition, the problem situations in these examples should still be relevant in the contemporary real world (which would not, for instance, be the case with examples containing societal statistical data or newspaper articles from the 1990s). Finally, 24 articles were selected plus an introductory theoretical contribution, altogether 25 chapters, a tribute to the anniversary. The authors were asked to slightly revise their articles by including recent experiences and/or by updating the problem situations. The book was eventually published in 2018 (Siller et al., 2018). In the following, we will briefly describe the content of this book.

The first chapter in the book (pp. 1–16) is the only one which was not selected from previous volumes but was written as a theoretical frame for the whole book. Werner Blum and Gabriele Kaiser, the two founders of the German Istron group, introduce theoretical approaches and empirical findings concerning mathematical modelling. Among other things, they discuss notions and aims of modelling as well as aspects of modelling competency, and they report on students' cognitive barriers and on appropriate teacher interventions.

The other 24 articles in the book are ordered chronologically, according to their original publication date in one of the Istron books, beginning with Heinz Böer's milk box example, published in 1993 in the first book of the series, and ending with

178 Cases of mathematical modelling

Hans-Stefan Siller's and Jürgen Maaß's sports bets example published in 2009 in the fifteenth book. We report here about the content of the book by grouping the articles according to the real-world contexts they deal with.

Most articles deal with the use of mathematics for understanding *everyday situations*. In the third chapter of the book (pp. 31–46), Michael Katzenbach asks whether families should travel to their holiday resorts by car or by train. The author developed a teaching unit for the lower secondary level (from grade 7 on) and reports on experiences with this unit. The mathematical models occurring in this example are mainly (stepwise) linear functions. Wilfried Herget analyses in the fourth chapter (pp. 47–68) various codes that are used to identify articles for sale, such as the ISBN or the GTIN (see Figure 7.18). The mathematics needed here is mostly elementary arithmetic including divisibility. The teaching unit developed by the author can be implemented from grade 6 on.

FIGURE 7.18 Example of a GTIN (Global Trade Item Number)

The eighth chapter in the book (pp. 111–124) deals with the question of what a positive result of an HIV/AIDS test means. Heinz Böer analyses the results of the underlying probabilistic models, which can be handled from grade 10 on. Regina Bruder presents, in the tenth chapter (pp. 133–144), various real-world situations within students' horizon of experience, which can be described by mathematical means, such as wrapping of sweets, shapes of tents, measures of school gardens, garbage fees or food prices. The author suggests various problem-solving strategies which can help develop students' modelling competencies. Hans Humenberger offers in the twelfth chapter (pp. 161–176) an elementary explanation of the socalled "Benford law". This law states that the first digit in numbers occurring in natural contexts (such as half-lives of radioactive substances, distances between places, measures of natural phenomena or numbers in tax declarations) is not uniformly distributed across 1, 2, . . . , 9 but these figures occur with decreasing frequencies. These empirical frequencies can be described amazingly well by certain logarithms (see Figure 7.19a).

An introduction to the concept of function in grade 7 by means of authentic diagrams is the topic of the fourteenth chapter (pp. 193–200). Johannes Schornstein reports on his own experiences as a teacher when using diagrams for the climate in a museum, the speed of a truck (see Figure 7.19b), the water consumption in a city or

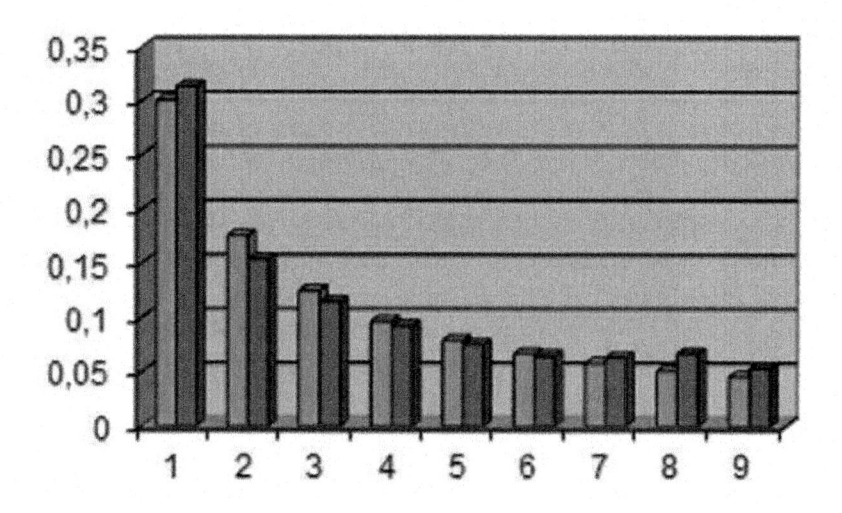

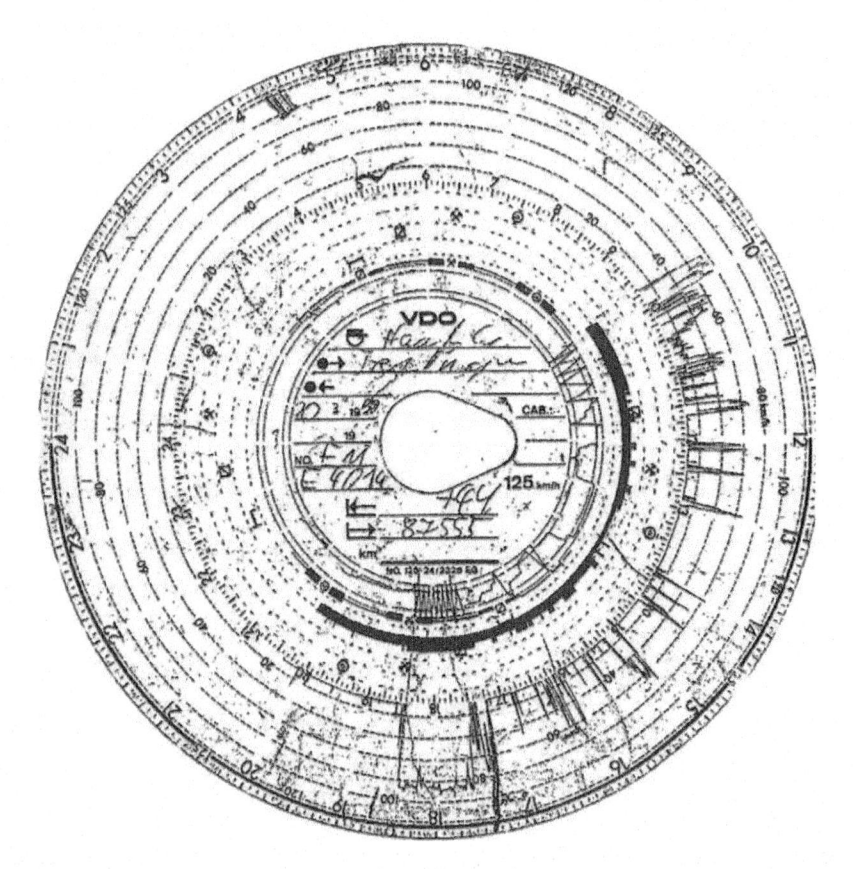

FIGURE 7.19a (top). Frequencies of first digit in a Google experiment (right columns) and according to Benford's law (left columns)

FIGURE 7.19b (bottom). Tachograph of a truck

the train timetable in a region. In the next chapter (pp. 201–230), Henning Körner presents teaching units in which students have discovered exponential functions for describing various real-world phenomena such as games, leading to geometric sequences, cooling of coffee, concentration of drugs in the blood, flow of liquids, the world population or the capacity of wind energy. The author concludes with the remark that, according to his experiences as a teacher, it is certainly possible to include modelling in everyday teaching. In the seventeenth chapter (pp. 245–260), Hans-Wolfgang Henn explains the everyday phenomenon of rainbows. The corresponding mathematical models, based on some physics such as the fraction of light, use real functions involving inverse trigonometric functions (see Figure 7.20). Since derivatives are needed for determining the maxima of these functions, this example is suited for the upper secondary level only.

Wilfried Herget and Dietmar Scholz analyse, in the nineteenth chapter (pp. 269–283), many mathematical aspects pertaining to newspapers. In particular, they have found many model mistakes in newspaper articles. The examples range from toilet paper over withers of calves, amount of precipitation, income of soccer stars and alcohol in the blood to education expenditure in Germany. In the twenty-first chapter (pp. 293–301), Gilbert Greefrath presents mathematical models of the filling of a home oil tank. The starting point is a newspaper article about fraud in the context of oil filling, and the aim of the teaching unit presented is to find ways of discovering such fraud by means of mathematics. In the following article (pp. 303–317), Timo Leuders deals with the question of why and how a dog seems to use the quickest path when it must bring back an object thrown into the water. The mathematical problem is the same kind of optimisation problem as is involved in understanding the paths of light beams.

Another group of articles consists of examples showing actual uses of mathematics in *professional contexts*. The second chapter in the book (pp. 17–30) contains one of the most classic Istron examples, the milk box, written by Heinz Böer. The question is which measures a one litre milk box, shaped as a right-angled parallelepiped with quadratic cross section and fulfilling certain practical requirements, must have if the consumption of surface material is to be minimal (see Figure 7.21). The corresponding mathematical model is a certain rational function which can be best analysed with mathematical tools from calculus, so the example is primarily suited for the upper secondary level. It turns out that the solution of this optimisation problem is very close to the measures of the milk box actually used in Germany. With this example, Böer won a modelling challenge in the early 1990s tendered by COMAP.

In the thirteenth chapter (pp. 177–191), Jörg Meyer analyses and clarifies some seeming paradoxes of descriptive statistics, which occur in the context of elections or school marks. The problems comprise questions such as what a majority vote, or the relation "better than", could mean. Also, the famous Simpson paradox, concerning the possible vanishing of sub-population trends under whole population aggregation, is discussed. Reinhard Oldenburg presents, in the sixteenth chapter (pp. 231–243), the mathematics of image processing. The mathematical topics

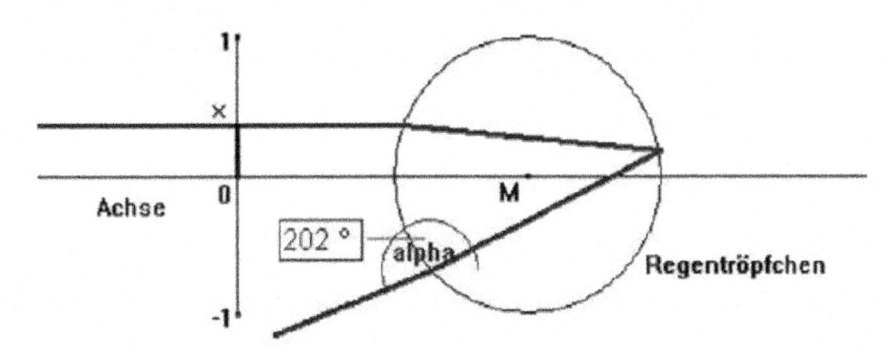

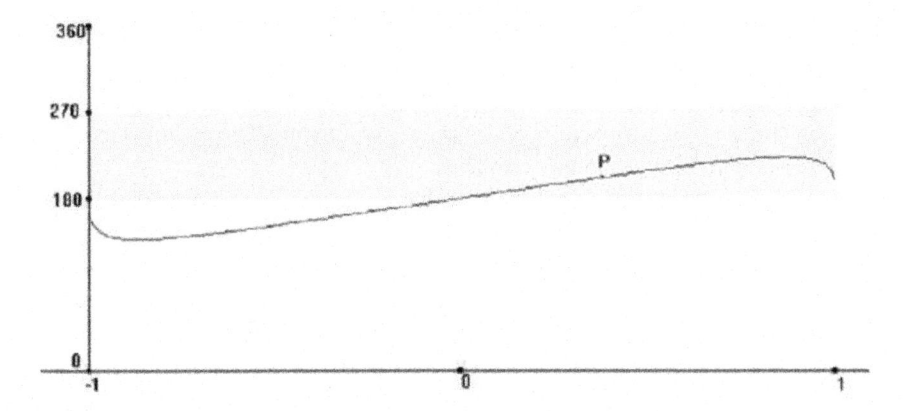

FIGURE 7.20 Rainbows and explanation of the main bow by means of third order beams

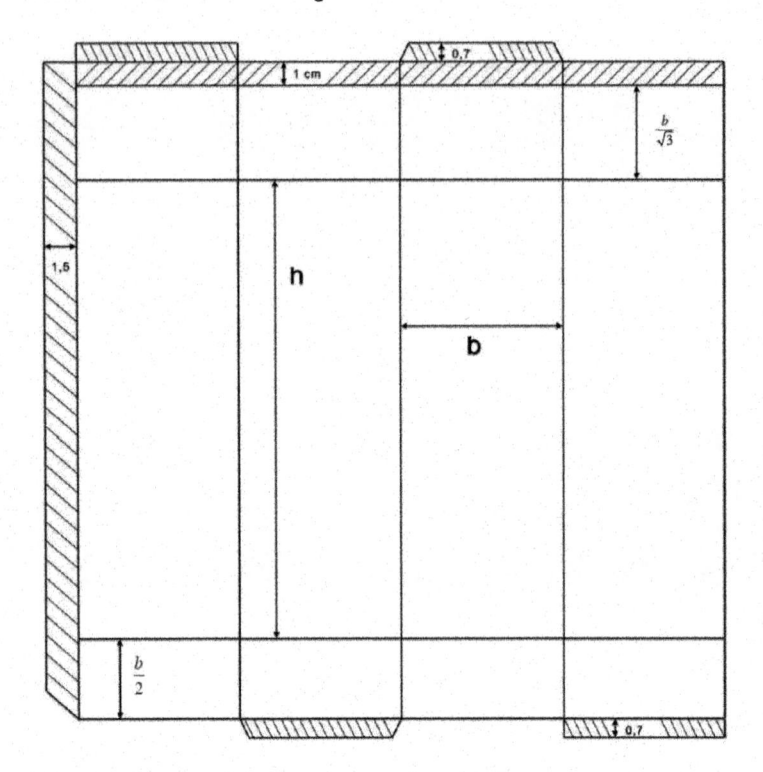

FIGURE 7.21 The surface of a milk box

involved include statistics, functions, geometric transformations and matrices, all of which are accessible at the upper secondary level. The final example uses Fourier transformations and exceeds the normal level of school mathematics. In the twenty-third chapter (pp. 319–330), Frank Förster presents the problem of determining the length of a cable wound around a drum with given measures. This problem is dealt with separately in the "cable drum" case in this chapter. In the next contribution (pp. 331–342), Christoph Ableitinger, Simone Göttlich and Thorsten Sickenberger deal with the problem of overbooking flights, using authentic data from an Austrian airline. The mathematical model is essentially a linear function of several variables, and the task is to find its maxima under certain constraints. The authors report on experiences with upper secondary students.

Yet another group of contributions is devoted to mathematical models in the context of *cars and traffic*. One instance is found in the fifth chapter in the book (pp. 69–78). Ingo Weidig analyses the mathematics behind the construction of mountain tracks for trains and presents a teaching unit for the lower secondary level with an existing track in the German Black Forest as an authentic example (see Figure 7.22). The mathematics involved includes percentages and elementary geometry up to circles.

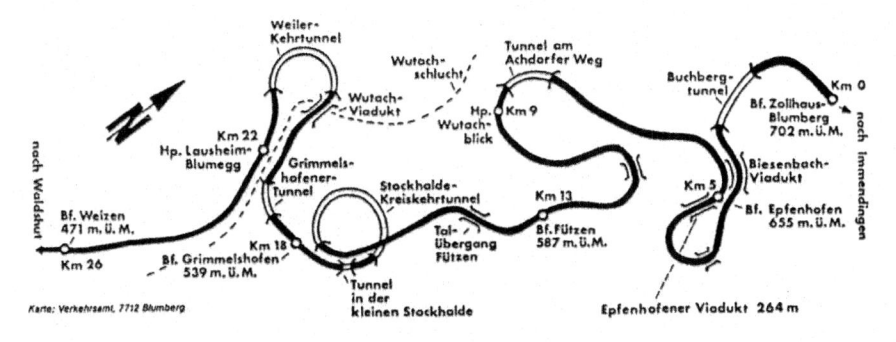

FIGURE 7.22 The track of the Wutachtal train in Germany

The ninth chapter (pp. 125–132) by Thomas Jahnke models the number of people in a traffic jam. The teaching unit developed by the author needs only elementary arithmetic and can be treated at the primary level already. In contrast, Hans-Wolfgang Henn, in the eleventh chapter (pp. 145–160), uses derivatives for determining speed and acceleration of a car and integrals for reconstructing speed from acceleration and distance from speed. This is suited for the upper secondary level in Germany. In the introductory teaching unit of calculus presented by the author, real data from a Porsche are used. In the eighteenth chapter (pp. 261–267), Jürgen Maaß presents a unit in which students in grade 9/10 developed a "radar speed trap". The core of the unit is students' own speed measurements, based on the usual formula for average speed. The twentieth chapter (pp. 285–292) contains a modelling project for the lower secondary level that the author, Katja Maaß, has developed and carried out in her own classes. The students were to determine the amount of paint needed to paint a Porsche. This problem is presented in more detail in the "Surface of a Porsche" case in this chapter.

Mathematical modelling in *sports* is a focus of three contributions in the book. Peter Bender analyses in the sixth chapter (pp. 79–96) the geometrical nature of soccer balls (see Figure 7.23), which in some versions are certain Archimedian polyhedra. The author shows how conditions of production influence the design of such balls.

FIGURE 7.23 Soccer balls

184 Cases of mathematical modelling

In the next chapter (pp. 97–109), Peter Bardy presents mathematical models of shot putting. An elementary quadratic model produces results which are too far away from real data, so the author constructs a more elaborate model consisting of a system of two differential equations, which has interesting consequences for the practice of shot putting. The twenty-fifth and final chapter in the book (pp. 343–356) has sports bets as its subject. Hans-Stefan Siller and Jürgen Maaß present a project in which students developed a model for bets on soccer matches. The mathematical content consists of percentages, proportions and elementary probability.

Altogether, the 25 contributions do indeed cover a broad spectrum of real-world contexts and mathematical topics. Teachers who can read German will find many sources for involving their students in genuine modelling activities.

Reference

Siller, H.-S., Greefrath, G. & Blum, W. (Eds.) (2018). *Neue Materialien für einen realitätsbezogenen Mathematikunterricht 4. 25 Jahre ISTRON-Gruppe – eine Best-of-Auswahl aus der ISTRON-Schriftenreihe*. Wiesbaden: Springer Spektrum.

Case 7: COMAP's *Mathematics: Modeling Our World*

In Chapter 4, we introduced the non-profit organisation Consortium for Mathematics and Its Applications, COMAP, established in 1980, as a protagonist in furthering applications, models and modelling in mathematics education in the USA. In this chapter, we zoom in on one of their many contributions to this area as a case in point. COMAP founded the so-called ARISE project (Application Reform in Secondary Education) in 1992 and obtained funding of it from the US National Science Foundation. The main offspring of the project was a four-volume set of curriculum materials for upper secondary education, *Mathematics: Modeling Our World*, consisting of textbooks and activities for students accompanied by commentaries and guidelines for teachers. The project was led by COMAP founder and director Solomon Garfunkel, high school teacher Landy Godbold and Columbia University professor Henry Pollak. The authorship of the four volumes was in the hands of modelling specialists, mathematics educators and high-profile practising high school mathematics teachers. The first book came out in 1998, published by South-Western Educational Publishing. The current – second – edition is published by COMAP itself (COMAP, 2000, 2010, 2011, 2013).

In the introduction to the material in the first volume, Sol Garfunkel wrote the following programmatic statement about ARISE (COMAP, 1998, T1):

> The result of these labors is *Mathematics: Modeling Our World*. In the COMAP Spirit, *Mathematics: Modeling our World* develops mathematical concepts in the contexts in which they are actually used. The word "modeling" is the key.

Real problems do not come at the end of chapters. Real problems don't look like mathematics problems. Real problems are messy. Real problems ask questions such as: How do we create computer animation? How do we effectively control an animal population? What is the best location of a fire station? What do we mean by "best"?

In this statement, mathematical concepts are justified by and arise out of attempts to model extra-mathematical problem situations. In other words, the statement insists on perceiving, in one grip, the whole rationale of the endeavour as "modelling for the sake of mathematics for the sake of modelling". This is indeed reflected in the structure and organisation of each of the four volumes, albeit less so in the fourth volume on pre-calculus. Each chapter in each book (again with a slight exception in the fourth volume; see below) is devoted to an extra-mathematical theme lending itself to mathematical modelling. The theme is being presented stepwise in a number (between two or six) of "lessons", each of which is supposed to be dealt with in up to six days of work. For each lesson, the material stipulates special "activities" and "individual work" for students to be undertaken during the lesson. The lessons in a chapter are concluded with one to two days of chapter review activities. The chapter finally contains a mathematical summary, a glossary (in the first two volumes) and a recommended "chapter project" in the first volume.

The eight chapters and corresponding lessons of the first volume (2nd edition) are the following:

1. *Pick a Winner: Decision Making in a Democracy* (Two current Election Models; Two Alternative Election Models; Chapter Project: Point Models);
2. *Secret Codes and the Power of Algebra* (Keeping Secrets, UGETGV EQFGU; Decoding; Cracking Codes; Illusive Codes; Matrix Methods; Chapter Project: Designing a Model for Coding);
3. *Scene From Above* (Changing Times; It's All a Matter of Scale; Shape, Size, and Area; Areas of Irregular-Shaped Regions; Chapter Project: San Francisco Wetlands);
4. *Prediction* (The Hip Bone's Connected; Variability; Linear Regression; Selecting and Refining Models; Chapter Project: Let the Bones Speak!);
5. *Animation/Special Effects* (Get Moving; Get to the Point, Escalating Motion; Calculator Animation; Fireworks; Chapter Project: Calculator Animation);
6. *Wildlife* (First Steps, First Moose Model; Multiplicative Growth; Second Moose Model; Final Moose Model; Chapter Project: Funding a College Education);
7. *Imperfect Testing* (A Sporting Chance; On One Condition; Nobody's Perfect; But I didn't Do It! Really!; Chapter Project: Failing Twice); and
8. *Testing 1, 2, 3* (Steroid Testing; Testing Models; Confirming the Model; Solving the Model: Tables and Graphs, Solving the Model: Symbolic Methods; Chapter Project: Pooling Three Samples).

186 Cases of mathematical modelling

The seven chapters and corresponding lessons of the second volume (2nd edition) are the following:

1. *Gridville* (In Case of Fire, Linear Village; Absolute Value; Minimax Village; Return to Gridville);
2. *Strategies* (Decisions; Changing Your Strategy; Changing the Payoffs; Optimal Strategies; Optimal Strategies Revisited; Games That Are Not Zero Sum);
3. *Hidden Connections* (Connections; Procedures; Minimum Spanning Tree Algorithms; Coloring to Avoid Conflicts; Traveling Salesperson Problems; Matching);
4. *The Right Stuff* (Packing Models, Designing a Package; Technological Solutions; Getting the Facts; Packaging Spheres);
5. *Proximity* (Colorado Needs Rain! Neighborhoods; Rainfall; A Method of a Different Color; Digging for Answers);
6. *Growth* (Growing Concerns; Double Trouble; Finding Time; Sum Kind of Growth; Mixed Growth); and
7. *Motion* (Learning Your Lines; Falling in Line; It Feels Like Fall; What Goes Up Must Come Down; The Grand Finale).

The third volume (2nd edition), too, contains seven chapters with corresponding lessons as follows:

1. *The Geometry of Art* (Keep It In Perspective; Drawn to Scale; Vanishing Point; The Right Space; The View From the Edge; Foreshortening);
2. *Fairness and Apportionment* (Heir Today, Gone Tomorrow; More Estate Division; Apportionment: The Unfairness of Fairness; Other Methods; Measuring Unfairness);
3. *Sampling* (It's All In the Question; Experience Counts; Say It With Confidence!; Selective Service; The Results Are In! Tag, You're It!);
4. *Mind Your Own Business* (So, You Want to Be in Business; Who's Minding the Store(room)?; Changing Assumptions; Slow Growth);
5. *Oscillation* (Life's Ups and Downs; A Sine of the Times; Connections; Fade Out; Now We're in Cookin');
6. *Feedback* (What Lies Ahead; Another Model; It's Going Around; An Ecological PushMe-PullYou); and
7. *Modeling Your World* (The Modeling Process; Analyzing Mathematical Models; Modeling Our World; Creating Your Model).

The fourth and final volume, titled *Pre-Calculus* (1st Edition), contains eight chapters and corresponding lessons as follows:

1. *Functions in Modeling* (A Theory-Driven Model; Building a Tool Kit of Functions; Expanding the Tool Kit of Functions; Transformations of Functions; Operations on Functions);

Cases of mathematical modelling **187**

2   *The Exponential and Logarithmic Functions* (Exponential Functions, Logarithmic Scale; Changing Bases; Logarithmic Functions; Modeling With Exponential and Logarithmic Functions; Composition and Inverses of Functions);
3   *Polynomial Models* (Modeling Falling Objects; The Merits of Polynomial Models; The Power of Polynomials; Zeroing in on Polynomials; Polynomial Divisions; Polynomial Approximations);
4   *Coordinate Systems and Vectors* (Polar Coordinates; Polar Form of Complex Numbers; The Geometry of Vectors; The Algebra of Vectors; Vector Equations in Two Dimensions; Vector Equations in Three Dimensions);
5   *Matrices* (Matrix Basics; The Multiplicative Inverse; Systems of Equations in Three Variables);
6   *Analytic Geometry* (Analytic Geometry and Loci: Modeling with Circles; Modeling with Parabolas; Modeling with Ellipses; Modeling the Hyperbolas);
7   *Counting and the Binomial Theorem* (Counting Basics; Compound Events; The Binomial Theorem); and
8   *Modeling Change With Discrete Dynamical Systems* (Modeling Change with Difference Equations; Approximating Change with Difference Equations; Numerical Solutions; Systems of Difference Equations).

The first three volumes are structured with extra-mathematical themes as the organising principle, whereas mathematical topics constitute the main organising principle in the fourth volume, which is because this book, by dealing with pre-calculus, delivers some degree of standard preparation for college studies for students in the last year of high school while still giving a prominent role to modelling situations and issues whenever possible.

In principle, the four books address the four years of high school in the USA at large. However, the project directors did not write and produce them with the primary aim of achieving high sales rates, and presumably they never expected them to be widely used in American high schools. Rather, they were produced as an existence proof of the possibility of actually basing an entire high school mathematics curriculum, covering all relevant mathematical topics, on the modelling of extra-mathematical situations. As expected, the books did not reach high sales – they were primarily adopted by schools in New York, Minnesota and Ohio. However, they are widely consulted as inspiration materials by teachers who teach according to more traditional materials. This is analogous to the fact that certain books in literary fiction and poetry that never reached a wide readership are, nevertheless, significant because they are reference works and provide inspiration to writers and poets in their own work.

Notes

1   Model M1a can easily be generalised from the special cable drum in figure 7.1 by using variables L, R_1, R_2 and r instead of concrete quantities, and the same is possible for the other models. The radii in model *M1a* are $R_2 + (2i - 1)r$ for i = 1, . . ., n where n is the number of layers fitting on the drum, that is, the largest integer below $(R_1 - R_2): 2r$. In each layer, there

are m cable pieces next to one another where m is the largest integer below $L:2r$. So the total length of all $m \cdot n$ pieces is:

$$C = \Sigma m \cdot 2\pi \cdot (R_2 + (2i - 1)r) = 2\pi m \cdot (nR_2 + r \cdot \Sigma(2i - 1)) = 2\pi m \cdot (nR_2 + n^2 \cdot r)$$
$$= 2\pi mn \cdot (R_2 + nr)$$

since the sum of all odd numbers up to $2n - 1$ *is* n^2.

2 The German Istron group was founded as part of an international network. The idea of this network was born at a meeting of a small international group in 1990 in the Istron Bay Hotel on Crete, Greece, hence the name of the group. The two authors of this book took part in this meeting.

References

COMAP (1998). *COMAP's Mathematics: Modeling Our World*. Annotated Teacher's Edition. Cincinnati, OH: South-Western Educational Publishing.

COMAP (2000). *COMAP's Mathematics: Modeling Our World* (Vol. 4 – Pre-Calculus, 1st Edition). Bedford, MA: COMAP, Inc.

COMAP (2010). *COMAP's Mathematics: Modeling Our World* (Vol. 1, 2nd Edition). Bedford, MA: COMAP, Inc.

COMAP (2011). *COMAP's Mathematics: Modeling Our World* (Vol. 2, 2nd Edition). Bedford, MA: COMAP, Inc.

COMAP (2013). *COMAP's Mathematics: Modeling Our World* (Vol. 3, 2nd Edition). Bedford, MA: COMAP, Inc.

8
FOCAL POINTS FOR THE FUTURE

8.1 The focal points and contributions of this book

One of the major ambitions for this book has been to provide the reader with a coherent up-to-date conceptual and theoretical framework about mathematical modelling in mathematics education. This means that the overarching focus has been the process of constructing and using mathematical models of extra-mathematical contexts and situations, in the context of teaching and learning of mathematics. Even though much of what has been said in the previous chapters pertains to all levels of mathematics education, from primary to tertiary, particular attention has been paid to school mathematics (ages 7–18), especially at the secondary level.

Against this background, we have offered (in Chapter 2) an extensive treatment of the fundamental general notions of "model" and "modelling" – descriptive and prescriptive. These are accompanied by a detailed exposition of seven modelling examples (Chapter 3), each of which is accessible at some level of school mathematics. This provided the platform for an analysis of the cognitive demands involved in undertaking mathematical modelling (section 2.6) and of what has become known under the heading of mathematical modelling competency and (sub-)competencies (Chapter 4). The possible places and roles of mathematical modelling in the education system, especially in schools, were discussed in Chapter 2 (section 2.8), also including a historical perspective (section 2.7). The most significant challenges to the implementation of modelling in the teaching and learning of mathematics were charted and analysed in Chapter 5.

An extensive account of empirical research on a multitude of aspects of what might be termed "the didactics of mathematical modelling" formed the core of Chapter 6, while a description and analysis of a number of selected cases of implementation of different sorts and at different levels was given in Chapter 7. In the next section, we shall take a retrospective look at key elements of what constitutes state-of-the-art knowledge and insights concerning mathematical modelling in mathematics education, thus summarising what was written in previous chapters.

8.2 What do we know from research and development, and what are we able to do?

First, it is important to note that the field is well under way in establishing a conceptual framework and a corresponding terminology about the didactics of mathematical modelling. Essential components of this framework are the very notions of model and modelling. Another key component is the modelling process, represented diagrammatically by various versions of the modelling cycle. This is to be understood as an analytic reconstruction of the steps necessarily involved in constructing a mathematical model but not as a depiction of the actual actions that a modeller must or will go through in any given modelling context. Further crucial components are: modelling competency and (sub-)competencies; model-eliciting activities; descriptive and prescriptive modelling; emergent modelling; modelling as a vehicle for other purposes versus modelling as a goal of mathematics education in its own right; and holistic versus atomistic approaches to the teaching and learning of modelling. While there exists no unified, universally agreed-upon conceptual and terminological framework with fixed definitions for the modelling discourse, there is, after all, a degree of consensus about these matters. This suffices for productive investigations and discussions to be undertaken across different quarters and groupings within the international mathematics education community.

Next, we know that the ability to successfully carry out mathematical modelling is by no means an automatic consequence of possessing a high level of intra-mathematical competencies, skills and knowledge. This implies that mathematical modelling must be learnt. Furthermore, the ability to perform mathematical modelling requires a minimum of mathematical knowledge and competence, but some students are able to successfully engage in modelling on a relatively sparse mathematical base. Fortunately, we also know from research and practice that mathematical modelling can be learnt as a result of high-quality goal-oriented teaching within carefully designed teaching-learning environments with sufficient structures, amounts of time and other resources made available to students and teachers. It is unlikely that modelling will be learnt by more than a few students if the education system is unable or unwilling to pay the necessary material and immaterial costs.

In this context, research has taught us quite a lot about why learning to model is demanding and difficult. We now know what it takes for students to successfully learn to model, as well as what barriers and obstacles exist to this endeavour. This has been studied from the perspectives of cognition, affect and beliefs about mathematics (see Chapters 5 and 6). We know that it takes a sustained and systematic effort over a long period of time to teach students to model. We also know that the conditions for achieving this may conflict with educational traditions, habits and conceptions. Learning to model does not fit the usual didactical contracts and socio-mathematical norms in mathematics classrooms, particularly those widespread conceptions of mathematics teaching focusing on training students to solve tasks of an algorithmic type which subsequently constitute the core of summative assessment and examinations. Last but certainly not least, we know that although the mathematics teacher is always a decisive factor in students' learning of mathematics, this is even truer as far as mathematical modelling is concerned. The teacher must be able and willing to leave his or her own comfort zone and, together with his or her students, to enter unexplored domains.

Regarding the role of digital technologies and tools in mathematical modelling (see section 5.7), we know from research and practice that such tools can greatly enhance the range and quality of many kinds of modelling activities in mathematics education. However, the use of the very same piece of technology can lead to marvels as well as disasters, depending on the didactico-pedagogical thinking and work of the teacher. The successful use of digital tools in modelling requires carefully designed and implemented modelling tasks and environments in which the division of labour between technology-based and other mathematical work is carefully balanced. We also know that digital tools can in no way replace modelling competency or (sub-)competencies. On the contrary, the stronger the tools, the stronger the need for modelling competency and mathematical competence at large.

When examining the extent to which students have successfully dealt with a given modelling task or made progress in developing their modelling competency, we must make use of a range of appropriate assessment modes and instruments (see section 5.6). During the last three decades, remarkable progress has been made in developing such modes and instruments and in putting them to use in special educational settings. However, as the use of these is demanding and time consuming, it has proved difficult to implement them in "ordinary" educational settings, especially on a large scale such as national or state-based assessments.

Finally, it is evident that the manifest inclusion of mathematical modelling in the curricula and practices of mathematics education is much more prevalent in some countries and places than in others. Countries such as Australia, Brazil, Denmark, Germany, the Netherlands, Spain, Sweden and the UK are (or were) protagonists in the field, but there are also active groups in countries such as Austria, China, Japan, Mexico, Portugal, Singapore, South Africa, and the USA. There are individual people and institutions in every country around the world who are "activists" of mathematical modelling in a concerted manner, but in several places their role in this respect is of a rather singular nature. The most remarkable thing is that despite all the progress reported here, most countries in the world do not pay particular attention to mathematical modelling in mathematics education.

8.3 What do we want to know and be able to do?

In the preceding section, we offered a brief general outline of what has been achieved so far in the didactics of mathematical modelling. In this section, we shall attempt to take stock of what we would further like to know, understand and be able to do in the future.

Let us begin by reiterating that mathematical modelling is difficult, demanding and time consuming and that the ability to undertake it successfully can only by developed and consolidated by doing it in wide variety of contexts and situations. This suggests that mathematical modelling is highly situated and dependent of the specific boundary conditions and circumstances of the individual modelling situation. This gives rise to the following overarching question (addressed in section 5.1): Knowing that it is possible to learn to model in some contexts and situations, to what extent is it possible for an individual to acquire a general modelling ability that goes across and beyond a multitude of diverse disciplines and fields of practice without having been trained to model in every one of these disciplines or fields? Or, differently put, to what extent can modelling abilities developed within some domains be transferred to and be activated in completely different domains? This question is related to several further questions: To what extent is a student's specific knowledge about a given extra-mathematical domain decisive for the student's ability to undertake modelling with regard to that domain? Are there characteristic differences in this respect among different extra-mathematical domains, such that some domains require more or deeper substantive knowledge and insights than do others? What is the impact of students' specific knowledge of and interest in given modelling contexts and situations for their willingness and ability to learn to model more broadly? We have germs of answers to these questions but nothing like complete answers.

This is also true of another related question. We know that it is often very difficult for students to recognise and utilise structural equivalence of models and modelling situations pertaining to different extra-mathematical contexts, even if they themselves have had modelling experiences from one or more of these contexts. So what does it take for students to become able to recognise and utilise such structural equivalences so as to transfer these experiences to new domains, contexts and situations? How can we foster and further this ability?

The finding that it is (sometimes) possible for students to learn to model relies on a number of successful teaching/learning cases, some of which are reported in the modelling literature and in Chapter 7. Just as in mathematics education at large, there is ample evidence that transplanting examples that proved successful in one environment and setting to new ones only seldom engenders the same degree of success as in the original case. How can we characterise successful instructional examples to uncover the conditions, factors and causes for the success observed, in the hope that (some of) these may be generalised, transferred and scaled to entirely new settings and circumstances? What transformations, transpositions and other measures can be made to successful cases to ensure new successes?

Moreover, even though we have several examples of successful instructional sequences in which students learnt to model to a satisfactory degree, we have virtually no studies that allow us to assess the long-term effect, over several months or years, of the learning that took place in such sequences. We badly need such long-term studies.

Finally, we want to know much more about the interrelationships and dependencies between, on the one hand, modelling competency and (sub-)competencies and, on the other hand, intra-mathematical knowledge and competencies. This is a huge question, the robust answers to which will have wide-ranging consequences not only for the future destiny of mathematical modelling in mathematics education but also for mathematics education at large.

8.4 Challenges for the future

The Discussion Document for the 14th ICMI Study on Applications and Modelling in Mathematics Education (Blum et al., 2002) listed a number of issues, which the ICMI Study attempted to address. The ones that specifically pertain to mathematical modelling, rather than to the field at large, asked the following questions:

> To what extent is applications and modelling competency transferable across areas and contexts? What teaching/learning experiences are needed or suitable to foster such transferability?
>
> *(p. 155)*

> What does research have to tell us about the significance of authenticity to students' acquisition and development of modelling competency?
>
> *(p. 160)*

> How can modelling ability and modelling competency be characterised, and how can it be developed over time?
>
> *(p. 160)*

> How can modelling in pre-service and in-service education be promoted?
>
> *(p. 161)*

> What would be an appropriate balance – terms of attention, time and effort – between applications and modelling activities and other mathematical activities in mathematics classrooms at different educational levels?
>
> *(p. 162)*

> What are appropriate pedagogical principles and strategies for the development of applications and modelling courses and their teaching? Are there different principles and strategies for different educational levels?
>
> *(p. 164)*

> What alternative assessment modes are available to teachers, institutions and educational systems that can capture the essential components of modelling competency, and what are the obstacles to their implementation?
>
> *(p. 165)*

> How should technology be used at different educational levels to effectively develop students' modelling abilities and to enrich students' experience of open-ended mathematical situations in applications and modelling?
>
> *(p. 167)*

These questions are stated in a very general form that, if taken at face value, call for final and exhaustive answers that leave nothing to be desired. The questions represent "demands at infinity". It doesn't make sense to ask, "[W]hat proportion of the distance to infinity is now being covered by progress in research and development since the publication in 2002 of the Discussion Document for the ICMI Study?" However, revisiting these issues today allows us to claim that, as shown in the chapters of this book, remarkable albeit varying progress has indeed been made with regard to all the issues cited. At the same time, these issues constitute ongoing challenges for research and development in mathematics education in the future.

Reference

Blum, W. et al. (2002). ICMI study 14: Applications and modelling in mathematics education: Discussion document. In: *Educational Studies in Mathematics* **51**(1/2), 149–171.

INDEX

ability 1, 24, 76, 77, 78, 79, 80, 81, 82, 113, 131, 134, 190, 192, 193
ACER (Australian Council for Educational Research) 171
activity-oriented learning environment 97
adaptive intervention 124, 128
Adding It Up 77
affine function 32
Ahmes papyrus 26
algorithm 191
amortisation 20, 22, 57
analytic reconstruction 13, 25, 190
annuity 57
anti-derivative 169
Anwendungsfähigkeit 82
A-paper (DIN) 50
applied mathematical problem solving 18, 171, 174
approaches to including modelling 98
arc 46, 68
area 19, 51, 70, 154, 159, 160, 161, 162
ARISE 184
"Army bus" task 117, 127
assessment 94, 98, 102, 133, 166, 168, 172, 173, 174, 191, 193
assumption(s) 10, 12, 14, 19, 31, 51, 59, 62, 77, 118, 162, 164
atomistic approach 29, 81, 83, 88, 132, 134, 190
attitude 79, 115, 164
authenticity 103, 127, *156,* 193
authentic problem 29, 31, 103, 127, 156, 182
axiom 5

bar code 8, 178
barrier(s) 3, 22, 27, 90, 92, 95, 115, 177, 190
BASE course 165, 166, 170
belief 24, 94, 115, 164, 190
Bertrand's paradox 67
Bildungsstandards 77
Blomhøj, M. 28, 77, 81, 83, 165, 166, 169, 170
Blum, W. 14, 16, 27, 29, 77, 79, 82, 90, 95, 98, 119, 128, 131, 177
BMI (Body Mass Index) 20, 22
Borromeo Ferri, R. 17, 95, 104, 122, 131
bottom-up definition 80, 81, 83, 87
Brand, S. 76, 79, 81, 83, 84, 132
Burkhardt, H. 24, 26, 91, 99, 102, 117, 174

cable drum 146, 155, 156, 157, 182, 187
CAD (Computer Aided Design) 162
calculus 65, 167, 169
CAS (Computer Algebra System) 104
cases 4, 27, 145, 146, 164, 174, 176, 192
challenge(s) 3, 90, 91, 189, 194
chord 67
circle 44, 67, 114, 147, 149, 150, 151, 152, 153, 154, 156, 157
circumference 47, 70, 157
classroom management 123
CO_2 balance 169
COACTIV project 130
Cockcroft report 174
coding 7
cognition 17, 22, 79, 87, 190
cognitive activation 124

cognitive apprenticeship 4
COMAP (Consortium for Mathematics and its Applications) 26, 27, 101, 180, 184
Common Assessment Task 76, 171
competence 78, 91
competency 3, 76, 77, 79, 80, 84, 85, 86, 87
competency flower 84, 85
composite function 8, 119
compound interest 61
conceptual framework 3, 189, 190
constant repayment 56
consumption rate 38
context 1, 6, 21, 22, 30, 78, 86, 112, 116, 189, 192
criteria for modelling tasks 104
cross-section 45, 147, 153, 154
cuboid 160, 161, 162, 164
curriculum 4, 25, 97, 145, 164, 171, 172, 173
curve-fitting 17, 19

data 17, 18, 19, 54, 97, 105, 161, 164, 168, 169, 183
debt 57
degree of coverage 86, 87, 133
de-mathematisation 9, 11, 21, 25, 31, 78, 167
density 19, 26, 70
derivative 65, 183
descriptive modelling 20, 22, 28, 189, 190
DGS (Dynamic Geometry System) 105
diameter 69
didactical contract 93, 96, 120, 125, 190
didactics 3, 189, 191
differentiable 66
differential and integral calculus 167
differential equation 167
digital tool 104, 191
directive teaching 128
disc 71
Discussion Document for the 14th ICMI Study 79, 193, 194
DISUM project 114, 118, 125, 126, 127, 128
Doerr, H. 29, 102, 135
drawing 126, 147, 161
dressed-up problem 29, 31, 103, 117
driving school rule 64

Eiffel Tower 44
emergent modelling 29
empirical research 3, 27, 105, 111, 116, 175
enactment of mathematics 86
equation 42, 60, 119

equilateral triangle 67, 154
evaluate model 12, 154
evaluation/evaluating 12, 21, 25, 31, 76, 78, 120, 156, 157, 167
examination 93, 98, 99, 173, 190
Excel 169
exponential function 18, 87, 120, 166, 187
extra-mathematical domain 1, 6, 22, 28, 78, 80, 192
extra-mathematical question 11, 23

Faro do Cabo de São Vicente 48
feedback 98, 101, 168
Fermi problem 19
few years gap 93
filling up task 37, 96, 114, 116, 118, 129
five-fold nature of mathematics 1
Flaubert, G. 31
Förster, F. 146, 147, 150, 157, 182
foot length 36, 99
formative assessment 98, 102
Frejd, P. 99, 118, 132, 134
Freudenthal, H. 29, 90, 95
fundamental mathematical modelling 11

GAIMME Report 101
Galbraith, P. 16, 76, 102, 133, 171, 172
Garfunkel, S. 26, 184
geometric representation 8
geometric series 60
Gini coefficient 21, 172
goals (justifications) for modelling 20, 28, 135, 166, 190
Gravemeijer, K. 29
Greefrath, G. 105
Greer, B. 31
Grundvorstellung 119
guided re-invention 29

Haines, C. 77, 79, 99, 133
helix 150, 152
heuristic strategy 126
holistic approach 29, 81, 83, 84, 88, 132, 134, 190
horizon 44
horizontal mathematisation 5
hyperbola 65

ICME (International Congress on Mathematical Education) 26
ICMI (International Commission on Mathematical Instruction) 27, 79, 193
ICTMA (International Conference on the Teaching of Mathematical Modelling and

Applications; International Community of Teachers of Mathematical Modelling and Applications) 27, 76, 79, 101
idealisation 14, 31, 71, 119
IMMC (International Mathematical Modelling Challenge) 102, 171
implemented anticipation 23, 24, 25, 119
implicit model 132
inequality 21, 40, 42
in-service education 95, 193
integral 160, 169, 183
interest rate 9, 57
interpretation/interpreting 78, 120, 162
intra-mathematical 31, 86, 190, 193
inversion 9
invertible 8, 9
irrational 74
Istron group 188
Istron group (Germany) 102, 176

Jankvist, U.T. 115, 117, 120
Jensen, T.H. 28, 77, 81, 83, 133

Kaiser, G. 29, 76, 79, 80, 81, 82, 83, *84*, 97, 132, 177
Kjeldsen, T.H. 165, 169, 170
KOM project 77, 79, 81, 83, 84, 86, 133

Language of Functions and Graphs, The 175
Leiß, D. 16, 23, 114, 116, 124
Lesh, R. 135
lesson plan 123
lighthouse task 48, 120, 122, 125
linear function 10, 13, 17, 32, 64, 85, 87, 115, 119, 129, 166, 178
linear regression 166
loan 22, 56
logarithmic function 18, 87

Maaß, K. 76, 82, 94, 104, 159, 161, 162, 163, 164, 183
mapping 7, 32
match stick problem 99
mathematical aids and tools competency 85
mathematical answer 11
mathematical communication competency 85, 86
mathematical domain 7, 78
mathematical literacy 175
mathematical model 6, 7, 160, 165
mathematical modelling 6, 7, 26, 28, 145, 165; competency 28, 78, 176
mathematical problem handling competency 78, 85

mathematical problem solving 17, 24, 30, 162
mathematical question 11, 23, 24, 82, 85
mathematical reasoning competency 85
mathematical representation competency 85
mathematical symbols and formalism competency 85
mathematical theory 1, 86
mathematical thinking competency 78, 85
mathematical treatment 5, 18, 21, 24, 30, 31, 167
mathematics for the sake of modelling 28, 185
mathematisation 7, 23, 31, 78, 119, 167
MatLab 167, 169
MCM (Mathematical Contest in Modeling) 102
meta-cognition 83, 126
meta-cognitive activation 126
model 6
model eliciting 29, 102, 135, 190
model *for* 22, 29
model formulation 24
Modeling Our World 184–185
modelling cycle 13, 15, 16, 17, 20, 21, 28, 29, 31, 78, 80, 82, 105, 116, 127, 167, 170, 190
modelling for the sake of mathematics 28, 185
modelling materials 101, 174, 177
modelling process 76, 81, 83, 147, 190
modelling route 121
modelling skills 77, 133
modelling (sub-)competencies 3, 27, 28, 76, 78, 79, 80, 83, 86, 87, 131, 145, 156, 165, 177, 189, 190, 191, 193
model *of* 22, 29
models and modelling perspective 102, 135
models as a vehicle 135
mountain climber rule 47
MultiMa project 124
multiple solutions 124

Niss, M. 1, 7, 15, 20, 23, 27, 77, 81, 86, 90, 98, 99, 103, 117, 119, 133, 165
normative modelling 20
number plates 8
numeracy through problem solving 174

OECD 113
one-to-one correspondence 8
Open University 76
operative-strategic teaching 128
Optimisation 20, 21, 180

Ormiston College 172
oscillation 18
over-parametrised 12

paper format 21, 22, 50
parallel 68, 72
parallelepiped 169, 161
parameters 12, 18, 77, 85
PCK (pedagogical content knowledge) 130
person identification number 8
perspectives of modelling 29
physical experiment 71
PISA (Programme for International Student Assessment) 113, 130, 174
Pollak, H. 26, 184
Polya, G. 126
polynomial function 18, 87, 187
Porsche 159, 160, 161, 162, 164, 183
power function 18, 87, 119, 166
pre-mathematisation 14, 18, 23, 31, 78, 167
prescriptive modelling 20, 22, 28, 189, 190
principle of minimum support 124
probability 25, 67, 178, 184
probability distribution 70, 87
problem-oriented 165, 170
problems with patterns and numbers 175
professional development 3, 95
project work 165, 166, 170
proportional 10, 41, 47, 51, 66, 115, 136, 162, 164
Pythagorean theorem 45, 49, 129, 150, 151

QCAA (Queensland Curriculum and Assessment Authority) 173
quadratic function 64
quality teaching 98, 122, 128, 190
Queensland Board of Senior Secondary School Studies 171

radius of action 86, 87, 133
random 67
RAND Panel 77
rate 10, 29, 169
rate of change 29, 135, 169
rating 133
ratio 29, 36, 54, 154
rational function 64, 180
real model 14, 28, 82, 162
real world 14, 16, 26, 80, 86, 156, 159, 162, 164, 176, 177, 178, 184
reciprocal function 66
reduced situation 14, 23
regression analysis 17, 18
relevance paradox 90
repayment 56
representation 6, 8

revolving door 113
Roskilde University 77, 165, 170

scaffolding 124
scaling 29, 37, 51, 161
Schoenfeld, A. 24, 116, 124, 126
Schukajlow, S. 124, 125
sense-making 30, 31, 116, 127
sequence 51
Shell Centre at Nottingham University 26, 99, 102, 174, 175, 176
shoe size 36
sight range task 43
similarity 51
simplification/simplifying 19, 31, 117, 162
simulation(s) 67, 97, 105, 106
SINUS project 128
situated cognition 87, 112, 115, 127
situation 21, 22, 78, 86, 112, 116, 166, 167, 178, 189, 192
situation model 16, 23, 116
SMSG (School Mathematics Study Group) 26
socio-mathematical norms 190
solution plan 126, 128, 129
specifying 77, 78, 85
speed 62, 117, 168, 183
statistical method 18
statistical significance 12
Stillman, G. 16, 23, 81, 83, 116, 171, 172
stochastic model 74
story problems 30
strategic intervention 125
strategy 24, 85, 116, 119, 126
street number 7
structural properties 12, 21
structural similarity 115, 192
substitute strategy 116
summative assessment 98, 133, 168, 174, 191
suspension of sense-making 31, 116, 127
Swan, M. 99, 102, 174
syllabus 97, 173
systematisation 15, 16

tangent line 44, 48, 73
taxi 10, 12, 14, 19, 85
teacher competencies 95, 130
teacher education 95, 102, 131, 193
teacher interventions 124, 125, 126, 128
teaching material 27, 101
technical level 86, 87, 133
technology 191, 193
test 77, 105, 114, 126, 127, 129, 162, 164
thought experiment 67, 156, 162
top-down definition 80, 81, 83, 87

traffic flow task 62, 97, 104, 105, 119
transfer 93, 112, 113, 114, 115, 162, 192, 193
translation/translating 6, 11, 14, 25, 52, 59, 78, 92, 162
Treffers, A. 5
triangle 44, 67, 154, 160, 161
trigonometry 18, 47, 87, 180
triple 7, 29

under-parametrised 12
Uwe Seeler 5

validate answers 12, 21, 78, 82, 156, 164
validation/validating 12, 21, 22, 25, 31, 76, 78, 120, 154, 155, 156, 167

Verschaffel, L. 26, 31, 116, 117
vertical mathematisation 5
Victoria Board of Curriculum and Assessment 171
visualisation 104, 105, 106
volition 79, 80, 87
volume 38, 153, 154, 156, 164

Wake, G. 174
weather forecasting 25
Weinert, F.E. 79, 80
word problem 25, 26, 29, 30, 164
working mathematically 120

Zone of Proximal Development 124